Determining Core Capabilities in Chemical and Biological Defense Science and Technology

Committee on Determining Core Capabilities in Chemical and Biological Defense Research and Development

Board on Chemical Sciences and Technology
Board on Life Sciences

Division on Earth and Life Studies

NATIONAL RESEARCH COUNCIL
OF THE NATIONAL ACADEMIES

THE NATIONAL ACADEMIES PRESS
Washington, D.C.
www.nap.edu

THE NATIONAL ACADEMIES PRESS 500 Fifth Street, NW Washington, DC 20001

NOTICE: The project that is the subject of this report was approved by the Governing Board of the National Research Council, whose members are drawn from the councils of the National Academy of Sciences, the National Academy of Engineering, and the Institute of Medicine. The members of the committee responsible for the report were chosen for their special competences and with regard for appropriate balance.

This study was supported by Contract No. HQ0034-10-D-0003, TO #5 between the National Academy of Sciences and US Department of Defense. Any opinions, findings, conclusions, or recommendations expressed in this publication are those of the author(s) and do not necessarily reflect the views of the organizations or agencies that provided support for the project.

International Standard Book Number-13: 978-0-309-26535-5
International Standard Book Number-10: 0-309-26535-5

Additional copies of this report are available for sale from the National Academies Press, 500 Fifth Street, NW, Keck 360, Washington, DC 20001; (800) 624-6242 or (202) 334-3313; http://www.nap.edu/.

Printed in the United States of America

THE NATIONAL ACADEMIES

Advisers to the Nation on Science, Engineering, and Medicine

The **National Academy of Sciences** is a private, nonprofit, self-perpetuating society of distinguished scholars engaged in scientific and engineering research, dedicated to the furtherance of science and technology and to their use for the general welfare. Upon the authority of the charter granted to it by the Congress in 1863, the Academy has a mandate that requires it to advise the federal government on scientific and technical matters. Dr. Ralph J. Cicerone is president of the National Academy of Sciences.

The **National Academy of Engineering** was established in 1964, under the charter of the National Academy of Sciences, as a parallel organization of outstanding engineers. It is autonomous in its administration and in the selection of its members, sharing with the National Academy of Sciences the responsibility for advising the federal government. The National Academy of Engineering also sponsors engineering programs aimed at meeting national needs, encourages education and research, and recognizes the superior achievements of engineers. Dr. Charles M. Vest is president of the National Academy of Engineering.

The **Institute of Medicine** was established in 1970 by the National Academy of Sciences to secure the services of eminent members of appropriate professions in the examination of policy matters pertaining to the health of the public. The Institute acts under the responsibility given to the National Academy of Sciences by its congressional charter to be an adviser to the federal government and, upon its own initiative, to identify issues of medical care, research, and education. Dr. Harvey V. Fineberg is president of the Institute of Medicine.

The **National Research Council** was organized by the National Academy of Sciences in 1916 to associate the broad community of science and technology with the Academy's purposes of furthering knowledge and advising the federal government. Functioning in accordance with general policies determined by the Academy, the Council has become the principal operating agency of both the National Academy of Sciences and the National Academy of Engineering in providing services to the government, the public, and the scientific and engineering communities. The Council is administered jointly by both Academies and the Institute of Medicine. Dr. Ralph J. Cicerone and Dr. Charles M. Vest are chair and vice chair, respectively, of the National Research Council.

www.national-academies.org

COMMITTEE ON DETERMINING CORE CAPABILITIES IN CHEMICAL AND BIOLOGICAL DEFENSE RESEARCH AND DEVELOPMENT

MIRIAM E. JOHN (*Chair*), Private Consultant; Vice President, Emeritus, Sandia National Laboratories, Livermore, CA
DAVID R. FRANZ, Former Commander, USAMRIID, Frederick, MD
JILL M. HRUBY, Sandia National Laboratories, Albuquerque, NM
ANNA JOHNSON-WINEGAR, retired, Office of the Secretary of Defense
CHARLES E. KOLB, Aerodyne Research Inc., Billerica, MA
C. RICK LYONS, Colorado State University, Fort Collins, CO
JON MOGFORD, Texas A&M University, College Station, TX
RANDALL S. MURCH, Virginia Polytechnic Institute and State University, Arlington, VA
DONALD PROSNITZ, Private Consultant, Livermore, CA
TOM SLEZAK, E.O. Lawrence Livermore National Laboratory, Livermore, CA
HENRY H. WILLIS, RAND Corporation, Pittsburgh, PA
PATRICK SCANNON, (*Consultant*) XOMA, Berkeley, CA
GEORGE M. WHITESIDES, (*Consultant*) Harvard University, Cambridge, MA

National Research Council Staff

Board on Chemical Sciences and Technology

DOROTHY ZOLANDZ, Director
DOUGLAS FRIEDMAN, Program Officer
KATHRYN HUGHES, Senior Program Officer
AMANDA CLINE, Administrative Assistant
SHEENA SIDDIQUI, Senior Program Associate
RACHEL YANCEY, Senior Program Assistant

Board on Life Sciences

FRAN SHARPLES, Director
INDIA HOOK-BARNARD, Senior Program Officer

Acknowledgment of Reviewers

This report has been reviewed in draft form by individuals chosen for their diverse perspectives and technical expertise, in accordance with procedures approved by the NRC's Report Review Committee. The purpose of this independent review is to provide candid and critical comments that will assist the institution in making its published report as sound as possible and to ensure that the report meets institutional standards for objectivity, evidence, and responsiveness to the study charge. The review comments and draft manuscript remain confidential to protect the integrity of the deliberative process.

We wish to thank the following individuals for their review of this report:

R. Stephen Berry, University of Chicago
W. Peter Cherry, Independent Consultant
Joseph M. DeSimone, University of North Carolina
Gerald L. Epstein, US Department of Homeland Security
Thomas G. Ksiazek, University of Texas Medical Branch
Stephen S. Morse, Columbia University

Although the reviewers listed above have provided many constructive comments and suggestions, they were not asked to endorse the conclusions or recommendations, nor did they see the final draft of the report before its release. The review of this report was overseen by

Ed Przybylowicz, Retired, Eastman Kodak Company, and **Robert Frosch,** Woods Hole Oceanographic Institution. Appointed by the National Research Council, they were responsible for making certain that an independent examination of this report was carried out in accordance with institutional procedures and that all review comments were carefully considered. Responsibility for the final content of this report rests entirely with the authoring committee and the institution.

Contents

Appendixes

Acronyms

AFRL	US Air Force Research Laboratory
ARIEM	US Army Research Institute of Environmental Medicine
ASARDA	Assistant Secretary of the Army for Research, Development, and Acquisition
ASD(NCB)	Assistant Secretary of Defense for Nuclear, Chemical, and Biological Defense Programs
AT&L	Acquisition, Technology, and Logistics
BARDA	Biomedical Advanced Research and Development Authority
BSL	Biological Safety Level
BW	Biological Weapon
CB	Chemical and Biological
CBD	Chemical and Biological Defense
CBDP	Chemical and Biological Defense Program
CBRN	Chemical, Biological, Radiological, and Nuclear
CBW	Chemical and Biological Weapons
CDC	Centers for Disease Control and Prevention
COCOM	US Combatant Command
CONOPS	Concept of Operations
CRO	Contract Research Organizations
CRP	Critical Reagents Program
CTR	Cooperative Threat Reduction

DARPA	Defense Advanced Research Projects Agency
DASD(CBD)	Deputy Assistant Secretary of Defense for Chemical and Biological Defense
DHS	Department of Homeland Security
DNA	Deoxyribonucleic Acid
DoD	US Department of Defense
DOE	US Department of Energy
DPG	US Army Dugway Proving Ground
DTRA	Defense Threat Reduction Agency
ECBC	Edgewood Chemical Biological Center
EFP	Explosively Formed Projectile
FDA	US Food and Drug Administration
FFRDC	Federally Funded Research and Development Center
FSU	Former Soviet Union
GEIS	Global Emerging Infections Surveillance and Response System
GPHIN	Global Public Health Intelligence Network
GLP	Good Laboratory Practices
GMP	Good Manufacturing Practices
HHS	Department of Health and Human Services
IC	Intelligence Community
IED	Improvised Explosive Device
IND	Investigational New Drug
IR&D	Internal Research and Development
ISR	Intelligence, Surveillance, and Reconnaissance
JCAD	Joint Chemical Agent Detector
JCS	Joint Chiefs of Staff
JPEO-CBD	Joint Program Executive Office for Chemical and Biological Defense
JPL	Joint Priority List
JPM	Joint Program Manager
JPO	Joint Program Office
JRO-CBRND	Joint Requirements Office for Chemical, Biological, Radiological, and Nuclear Defense
JSIG	Joint Service Integration Group
JSMG	Joint Service Materiel Group

JSTO-CBD	Joint Science and Technology Office for Chemical and Biological Defense
LANL	Los Alamos National Laboratory
LLNL	Lawrence Livermore National Laboratory
MCMI	Medical Countermeasures Initiative
MDD	Material Development Decision
MRMC	US Army Medical Research and Materiel Command
NAS	National Academy of Sciences
NBACC	National Biodefense Analysis and Countermeasures Center
NCAR	National Center for Atmospheric Research
NCI	National Cancer Institute
NIH	National Institutes of Health
NIST	National Institute of Standards and Technology
NRC	National Research Council
NSRDEC	US Army Natick Soldier Research, Development, and Engineering Center
NSWC	Naval Surface Warfare Center
NTA	Non-Traditional Agent
OASD(NCB/CB)	Office of the Assistant Secretary of Defense for Nuclear, Chemical, and Biological Defense Programs/Chemical and Biological Defense
OGA	Other Government Agency
OSD	Office of the Secretary of Defense
PET	Positron Emission Tomography
PI	Principal Investigator
PNNL	Pacific Northwest National Laboratory
POM	Program Objective Memorandum
R&D	Research and Development
R-D-A	Research-Development-Acquisition
RDT&E	Research, Development, Test, and Evaluation
RNA	Ribonucleic Acid
RSDL	Reactive Skin Decontamination Lotion
S&T	Science and Technology
SNL	Sandia National Laboratory

T&E	Test and Evaluation
TMT	Transformational Medical Technologies
TMTI	Transformational Medical Technologies Initiative
US	United States
USAMRICD	US Army Medical Research Institute of Chemical Defense
USAMRIID	US Army Medical Research Institute of Infectious Diseases
USD	Under Secretary of Defense
USDA	US Department of Agriculture
USG	United States Government
USSOCOM	US Special Operations Command
UTMB	University of Texas Medical Branch
WDTC	West Desert Test Center
WRAIR	Walter Reed Army Institute of Research

Summary

The goal of the US Department of Defense's (DoD's) Chemical and Biological Defense Program (CBDP) is to provide "support and world-class capabilities enabling the US Armed Forces to fight and win decisively in chemical, biological, radiological, and nuclear environments."[1] To accomplish this objective, the CBDP must maintain robust science and technology capabilities to support the research, development, test, and evaluation required for the creation and validation of the products the program supplies to the Services. As the threat from chemical and biological attack is an evolving one, due to the changing nature of conflict and rapid advances in science and technology, the core science and technology (S&T) capabilities that must be maintained by the CBDP must also continue to evolve. In order to address the challenges facing DoD, the Deputy Assistant Secretary of Defense for Chemical and Biological Defense (DASD(CBD)) asked the National Research Council (NRC) of the National Academy of Sciences to conduct a study to identify the core capabilities in science and technology that must be supported by the program.

The NRC Committee on Determining Core Capabilities in Chemical and Biological Defense Research and Development has examined the capabilities necessary for the chemical and biological defense science and

[1] US Department of Defense. 2010. "Department of Defense Chemical Biological Defense Program Annual Report to Congress 2010." Note that this report focuses only on the chemical and biological aspects of the program.

technology program in the context of the threat and of the program's stated mission and priorities. This report contains the committee's findings and recommendations. It is intended to assist the DASD(CBD) in determining the best strategy for acquiring, developing, and/or maintaining the needed capabilities.

Because science and technology development is a long process, the products and materials from the CBDP must not only respond to the needs of the Services today, but also anticipate those of the future. Since the United States cancelled its offensive program for biological weapons in 1969 and for chemical weapons a decade later,[2] DoD must rely on analysis and simulations to understand how these agents might be used. Offensive programs are being conducted by adversaries where understanding of their intentions and capabilities is uncertain; at least some of the technology developed by nation states has escaped, and capability in all parts of the world in civilian uses of biotechnology and medicinal chemistry is rising rapidly. Because of uncertainty about how much protection current materiel and procedures will provide, there is potential for a gap between needs and deployable capabilities. For example,

- Do we need new, more effective vaccines?
- Does the current protective gear adequately protect, and against what agents?
- Would warfighters be able to "fight through" operations that use conventional agents (for example, persistent nerve agents) deployed in conventional ways? In innovative ways (for example, on suicide bombers)? With unconventional agents?
- How much would operational tempo be slowed by attacks on logistics and supply chains using chemical or biological weapons, and what would be the influence of successful attacks on operational tempo?

Many of the questions related to the capabilities that warfighters and combatant commands have at their disposal could be answered, but the current technical and organizational structure is not designed to answer them. Instead, the process for prioritizing research efforts and allocating resources is based on a requirements-driven process that promotes a focus on the development of technical solutions without adequately considering the range of contexts in which they may be used. This focus carries over into the approaches taken in evaluating the efficacy of the products. Although a device or material meets its threshold and objective bench-

[2] In 1969, the United States renounced first use of chemical weapons, and in 1991 renounced retaliatory use as well; the United States ratified the Chemical Weapons Convention in 1993.

marks in a test chamber or facility, these benchmarks may not necessarily represent the range of operating environments.

The committee identified 39 core chemical and biological defense S&T capabilities and created a framework that groups them in six categories. In Chapter 3, the committee discusses these S&T capabilities and identifies where, in their view, the capabilities should be obtained by the CBDP. To inform their thinking about which S&T capabilities are actually *core* and comment on where the capabilities may be found the committee developed a decision tree (Figure 3.1). Using this decision framework the committee found that almost all of the capabilities *can* be found outside of the service laboratories. The committee went on to identify, for a variety of possible reasons, some capabilities that *should* be maintained within DoD service laboratory infrastructure. For each capability, research and development (R&D) and test and evaluation (T&E) are discussed separately and typically were not best suited to the same organization.

The committee considered four types of institutions with laboratories that may be suited to provide CBDP core capabilities and organized them from typically having the most fundamental-science-focused to the most product-focused research. These institutions are (1) academia, (2) other governmental facilities (e.g., the National Institutes of Health, Centers for Disease Control and Prevention, Department of Energy National Labs, National Institute of Standards and Technology), (3) DoD laboratories and facilities, and (4) industry (e.g., pharmaceutical companies). For some capabilities, T&E requires use of actual agent;[3] institutions other than DoD laboratories may be well suited to do the work but would need to do so in close collaboration with DoD.

Table S.1 summarizes the committee's judgments about how well suited the types of institutions are for R&D and for T&E with respect to 26 of the core capabilities. Dark shades indicate an institutional category that the committee views as well suited to maintain a given capability for the CBDP, while the lighter shade indicates less well-suited locales. The white boxes indicate that the institutional category is, in the committee's view, not well suited to maintain the capability. The other 13 capabilities are cross-cutting science and technology that the committee views as necessary for effective RDT&E for any of the capabilities defined in the preceding capability categories. Discussion of the potential locales for the cross-cutting science and technology capabilities can be found in Chapter 3. The committee does not intend to imply that each of the 13 cross-cutting capabilities be maintained exclusively, or indeed at all, within DoD.

[3] Actual agent testing refers to the actual chemical or biological agent the capability is being tested against (e.g., Vx, Sarin, sulfur mustard, anthrax, tularemia, botulinum toxin), as opposed to testing with simulants.

Table S.1: Science & Technology Core Capabilities	R&D				T&E			
	Academia	OGF	DoD	Industry	Academia	OGF	DoD	Industry
1. Enabling CBRN ISR								
Information Acquisition & Analysis								
Health Monitoring								
Environmental Monitoring								
Unknown Agent Identification and Characterization								
2. Chemical & Biological Agent Detection								
Analytical Methods Discovery							A	
Instrumentation Development							A	
Sensor Systems Development							A	
Agent Transport Analysis							A	
3. Individual & Collective Protection								
Controlled Molecular Transport Materials Discovery							A	
Barrier Materials Engineering							A	
Personal Protective Systems Development							A	
Collective Protective Systems Development							A	
Physiology							A	
4. Medical Countermeasures								
Target Discovery								
Regulatory Science								
Mechanisms of Delivery & Delivery Systems								
Animal Models						N/A		
Host Response								
Pre-Clinical Studies		N/A						
Clinical Trials (GLP and GMP)								
Medical Product Development								
5. Hazard Assessment, Mgmt and Decon								
Decontamination Methods Discovery							A	
Decontaminant Development							A	
Decontamination Resilient Materials Development							A	
Decontamination Systems Engineering							A	
Agent Transport & Viability Analysis							A	

6. Cross Cutting Science and Technology
Acquisition, Maintenance and Transport of Critical Chemical and Biological Reagents
Simulation
Informatics
Forensics
Education and Training
Behavioral Analysis
Systems Analysis and Engineering
Repurposing Commercial Technologies
Systems Biology
Synthetic Chemistry and Biology
Materials Science
Statistical Measurement Design
Test & Evaluation

In the committee's view, this category is:
- not well suited
- less well suited
- well suited
- very well suited

to provide this capability to CBDP

R&D: research and development
T&E: test and evaluation
OGF: other govenment facilities
A: the actual chemical or biological agent is important for T&E
N/A: the capability does not have a major R&D or T&E component

When considering the various locales for obtaining S&T capabilities, it is important to recognize that

1. the analysis of the various laboratory locales is general, and individual performers within a category may be exceptions;
2. the color coding of each category represents the aggregate of reasons considered, including but not limited to
   a. reputation and experience at providing the given capability,
   b. the extent to which the capability requires work with classified information,
   c. limitations on the locale of the capability resulting from international treaties or other laws,
   d. the need to maintain important capabilities, at least in part, at government facilities to ensure availability (e.g., Biological Safety Level 4 facilities).

The committee identified a number of concerns that affect the program's ability to sustain these core capabilities, including

- the amorphous and changing nature of the threat;
- the breadth of the mission and lack of shared strategic objectives across all of the chemical and biological defense enterprise elements;
- a requirements-driven, as opposed to capabilities-based, process for prioritizing and directing RDT&E and acquisition;
- a funding structure that minimizes local flexibility over allocated RDT&E funds; and
- challenges to effective engagement with individuals and organizations external to DoD.

RDT&E for the CBDP rely upon capabilities that have been primarily resident in the military departments because of both the classified nature of the original offensive program and specialized aspects of the problem. While key competencies and special facilities in the laboratories and test ranges remain important to the program, most of the expertise in relevant science and engineering now lies outside of DoD. The work that *is* done largely by the military (for example, protective suits) is not carried out in a way that allows its effectiveness to be evaluated usefully, and the transfer of commercial technology into DoD laboratories (e.g., in gene sequencing) is inefficient and expensive. The fundamental questions of how RDT&E programs should be organized, and how much of chemical and biological defense research is really "core" to DoD require rethinking.

FINDINGS AND RECOMMENDATIONS

Program Framework and Structure

Mission and Strategy

Finding 1.1: The threat is unpredictable, changing, and dependent on the nature of conflict. The CBDP cannot rely on breakthroughs in intelligence on adversaries' chemical or biological terrorism or warfare programs to inform how its investments are prioritized.

Finding 1.2: The program has not adapted to the changing nature of the chemical and biological threat. It is impossible technically—and unfeasible economically—to try to provide solutions to all potential threats. The United States simply cannot afford to deal with all threats on an individual basis, and there is no universal solution—it has to choose which problems to solve.

Finding 2.1: The CBDP mission is too broadly stated. The stated mission of CBDP is to "Provide global chemical, biological, radiological, and nuclear defense capabilities in support of National Strategies." The mission statement is large enough to allow for a wide variety of interpretations, making it challenging for both the customers of the program and the facilities that support its work to understand the program priorities.

The CBDP has responsibilities that span missions from protecting the warfighter and providing support to the warfighter, to defending the United States from attack (i.e., Homeland Defense) and supporting local authorities following a chemically or biologically related incident (i.e., Consequence Management, Foreign or Domestic). Events requiring the Department to perform each of these missions could unfold in innumerable, unexpected ways; for example,

- naturally occurring disease or unintentional chemical exposures may be difficult to distinguish from intentional attacks;
- intelligence, as noted above, has historically proved uncertain and/or unreliable in assessing the chemical and biological (CB) threat;
- innovations, as witnessed in the case of improvised explosive devices, will certainly occur in the development and use of chemical and biological weapons; and
- the United States has a poor understanding of the intentions of those who might use chemical and biological weapons, and an even poorer understanding of the barriers that prevent them from doing so.

The combination of a broadly stated mission and numerous uncertainties calls for top-level guidance on focus and priorities, which the current program lacks.

Finding 2.2: There is no program-wide CB defense strategy, nor common characterization of the program elements among the participating organizations. The many different organizations currently involved in the CBDP (e.g., Office of the Assistant Secretary of Defense for Nuclear, Chemical, and Biological Defense Programs/Chemical and Biological Defense (OASD(NCB/CB), Joint Requirements Office for Chemical, Biological, Radiological, and Nuclear Defense (JRO-CBRND), Joint Science and Technology Office for Chemical and Biological Defense (JSTO-CBD), and Joint Program Executive Office for Chemical and Biological Defense (JPEO-CBD) each view the chemical and biological defense mission from different perspectives. Although this can be expected based on their different roles, the coordination and collaboration between these groups is far from seamless. As a result, the program has been—and continues to be—limited in its ability to deliver fielded solutions in a timely manner.

Finding 2.3: Strategic priorities tend to change with changes in senior leadership. As a result, efforts requiring sustained and/or longer-term commitments (e.g., medical countermeasures) are unable to deliver timely results, if at all.

Recommendation 2.1: The DASD(CBD) should lead a mission and strategy development activity that aligns all of the program elements and offices. The differences among offices in how they portray and communicate their stories, in their priorities, and in the terminology they use to describe the program are stark. Bold moves are needed to break the current stagnation that permeates the chemical and biological S&T and acquisition environment. Tweaking the management or refocusing a few projects will not be sufficient. The recommended alignment activity should promote a shared understanding of and commitment to key priorities for maintaining the core capabilities and expertise needed to fulfill the overall program mission and strategy.

Science and Technology

Scientific Collaboration

Finding 3.1: Little of the *fundamental* science required for CBD lies *primarily* in the DoD. The vast majority of the scientific research performed in the United States occurs in academic and industrial laboratories. This is particularly true for the biological and chemical sciences which lie at the nexus of the S&T requirements of the CBDP.

Finding 3.2: The military laboratory community is not as strongly partnered with key external research institutions and programs as it could and should be. As the United States has a robust S&T sector, the CBDP can and does engage with individuals and organizations external to DoD and the US government, but this typically occurs at the individual project or principal investigator level, and not necessarily on a sustained basis. The CBDP has not systematically promoted institutional ties with academic, industrial (especially pharmaceutical companies), and other non-DoD laboratories or related federal programs.

Recommendation 3.1: The Director, JSTO-CBD, should ensure that the development of a Culture of Collaboration is a high priority for all elements of the chemical and biological defense enterprise. Although information control requirements and contracting concerns have been stated as barriers on both sides to collaboration, these are issues that can and should be addressed. To ensure that the program delivers products based on the best S&T available, the CBDP needs to find ways to partner with the broader scientific community and other federal agencies in areas relevant to chemical and biological defense.

Tech Watch and Adopt

Finding 3.3: There is the potential to significantly improve chemical and biological defense capabilities by using existing technology. Despite the nation's superb biomedical research establishment and the explosive growth of biological and biomedical science that is relevant to DoD as well as the public health community, relatively little of this broad competency has been applied to problems relevant to chemical and biological defense.

Recommendation 3.2: The DASD(CBD) should establish an effective "tech watch and adopt" component within the CBDP to bring innovative solutions to ongoing needs. Program managers and scientists within the CBDP should recognize the importance of technology watch and adoption before a major new RDT&E investment is made. The incorporation of a "tech watch and adopt" concept would have at least the following three elements: (1) mechanisms for searching and identifying relevant breakthroughs in the literature and from the private sector; (2) mechanisms and processes in place for incorporating innovation into the ongoing program for the capability needed; and (3) processes for rapid adoption of "tweaks" that would significantly improve existing capabilities. An adjunct objective would be to get the external performers interested in CBD problems such that they might be recruited to work on the problem.

Linking R&D Community to Operators

Finding 3.4: Separation of S&T performers from the end user is impeding their ability to meet the user's needs. Individuals in the military laboratories noted that understanding more fully the context of their work could assist S&T personnel in developing operationally relevant products, identifying variables or factors that would otherwise be overlooked, and possibly shortening development time. In addition, a stronger relationship between operators and R&D performers could support innovation by enabling informed, collaborative "blue sky thinking."

Recommendation 3.3: The DASD(CBD) should survey the military laboratories and associated facilities to identify strong relationships between S&T performers and the warfighters, and support replication of such interactions across the program.

Simulants for Test and Evaluation

Finding 3.5: Broadly speaking, the *capacity* for test and evaluation to support the needs of the CBDP exists within DoD. Test and evaluation is a core component of the program and important to maintain within DoD at a high level of competency and responsiveness.

Finding 3.6: Much of the current T&E is based on *unrealistic* expectations of how the material or equipment being tested would *actually* be used. The threat, although long-standing, is uncertain. In addition, the lack of connection with the military operators often leads to the omission of realistic simulation of deployment and use environments.

Recommendation 3.4: Because of the economic, logistical, and environmental concerns with actual agent testing, DASD(CBD) should give priority to the active development and production of realistic and relevant threat agent simulants for both outdoor and large-chamber tests. A single simulant, especially for chemical agents, is unlikely to possess all of the same physical, chemical, and/or transport properties of an actual agent; therefore, multiple simulants may be required to fully stress critical design parameters during T&E.

Review of Test and Evaluation Plans

Finding 3.7: Test and evaluation plans apparently are not subject to independent external review. These plans are created internally, and the committee observed little evidence of the use of external expertise to review testing plans.

Recommendation 3.5: For CBD products to be viable for fielding, the Deputy Under Secretary of the Army for Test and Evaluation should require that (1) T&E activities be based on testing protocols that accurately emulate actual operating environments (both threat properties and operator employment) and (2) independent reviews of testing protocols be conducted.

Organization and Management

Capabilities-Based Planning, Development, and Acquisition

Finding 4.1: A requirements-driven S&T process is not a good match for the CBDP. The planning and experimentation carried out by the CBDP is usually so removed from plausible use that it is difficult to believe that the Combatant Commands would know how to understand and evaluate the program's impact, how best to protect their forces, to carry out their operations in the face of current and/or high-probability future threats. Planning tends to focus on narrow conceptions of threats and responses derived from historical events. Outcomes tend to be described in terms of consequences which can be easily measured, such as fatalities and injuries. Options tend to be developed based on incremental modifications to current materiel and operations. Each of these approaches is inadequate for addressing the evolving and innovative nature of chemical and biological threats. Moreover, the perceived goal of "100% protection" appears to impact all aspects of the program such that few products reach the field in a timely manner, especially in the medical countermeasures part of the program.

Recommendation 4.1: The Office of the Secretary of Defense (through the Assistant Secretary of Defense for Nuclear, Chemical, and Biological Defense Programs) should evaluate a shift to capabilities-based planning, as a more appropriate approach for this program. The goal is to adopt strategies that are *flexible* to provide capabilities for events other than those anticipated, *adaptive* to conditions other than those that are planned, and *robust* to attempts made to diminish these capabilities. Planning should expand the range of options considered; iterative review and realistic red-teaming should challenge assumptions built into plans and promote innovations in defense to correspond to that in the threats.

The scope of red-teaming and review should encompass the threats and activities against which performance is assessed and the evaluations of performance are made. The overall S&T focus should shift from "zero casualties" to "mission success."

Program Management

Finding 5.1: Successful transition between the JSTO-CBD and the JPEO-CBD offices requires a mutual agreement on appropriate transition points, encoded in multiyear program plans and budgets. Regardless of the chosen trigger, expertise and resources within or contracted by JSTO-CBD and JPEO-CBD need to be appropriately positioned. This approach would also be supportive of overlap in JSTO-CBD and JPEO-CBD personnel engagement on the project to ensure smooth and knowledgeable transitions. However, the committee observed that the partnership between the JSTO-CBD and JPEO-CBD is weak and that neither office viewed transition plans as a responsibility.

Finding 5.2: There is no end-to-end authority for the CBDP, which is particularly problematic for medical products. Though both JSTO-CBD and JPEO-CBD are overseen by the CBDP, there is no one office or individual with the responsibility and authority for the entire process for any given product. The risk—and reality—is that a transition gap between R&D and acquisition could result in the development of a project management "valley of death." The existing research-development-acquisition process may be adequate for acquiring the non-medical products in the CBDP. For the medical countermeasures program, however, FDA regulatory requirements must be considered early enough to influence product development decisions. The current management structure within the CBDP is not well suited to the task because of the lack of a whole-process, integrated view of product development.

Recommendation 5.1: The DASD(CBD) should evaluate alternative program management approaches, including incorporation of an end-to-end project management authority, especially for the medical countermeasures program.

Laboratory and Major Facility Management

Finding 5.3: The principal RDT&E military organizations associated with the CBDP are benefiting from major facility investments that are planned to provide both capabilities and capacities to meet the anticipated needs of the program. Operating and maintaining these facilities, however, will place a burden on both the owning Service (principally the Army) and the program. The initial operating plans appear to be resourced.

Finding 5.4: All or part of the elements required for healthy RDT&E activities were missing at the organizations visited by the committee. A successful RDT&E enterprise should include the following elements to ensure clarity of purpose, focus of investments, and coherence of management:

1. Clear mission and objectives
2. Continuity in leadership
3. The ability to understand, accept, and manage risk throughout the process
4. Predictable and stable funding
5. Effective asset management at the laboratory level
6. A sense of excitement and pride in the work among the staff

Of special concern are strained relationships between JSTO-CBD and the laboratories, the new rotational policy for military commanders in the Army, and a trend toward increasing oversight of both technical work and operations at the facilities.

Recommendation 5.2: The DASD(CBD) should formally review alternative laboratory management models, taking advantage of the numerous prior studies, reviews, and evaluations of laboratory and large facility management of S&T organizations. A principal objective is to define the level of stewardship that the program should provide to the principal RDT&E in-house facilities and laboratories.

Scientific Peer Review

Finding 5.5: All programs benefit from scientific peer review when done well, and these reviews keep the skills of scientists and engineers sharp.

Recommendation 5.3: The DASD(CBD) should implement a nested review process for chemical and biological defense RDT&E bound by consistent standards of rigor, frequency, and reporting. The CBDP and its supporting laboratories would each benefit from independent, periodic review at the programmatic and scientific levels. The CBDP should also encourage and participate in institutional reviews. An annual roll-up of review outcomes could help identify thematic areas of promise and concern.

1

Introduction

The US Department of Defense (DoD) has a long history in the area of chemical and biological defense (CBD). Over the course of that history, the program has evolved to address a broad spectrum of threats that would not and could not have been envisioned in the beginning. A critical part of the endeavor of DoD has been to engage smart and talented individuals to perform work in this area. The work has evolved over the years to what it is today. The program is at a turning point in terms of how it is able to implement its research, development, test, and evaluation (RDT&E) efforts to support the mission of the department.

To that end, the National Research Council was asked by the Office of the Assistant Secretary of Defense for Nuclear, Chemical, and Biological Defense to form an ad hoc committee that is able to address the current state of the enterprise and provide input into framing the program's efforts as it moves forward.

STATEMENT OF TASK AND PURPOSE OF STUDY

Statement of Task

This study was carried out in response to the following statement of task:

> The National Research Council (NRC) of the National Academy of Sciences (NAS) will identify the scientific and technology capabilities that must be available to support Chemical and Biological Defense Program

(CBDP) research, development, test and evaluation, and operational activities, and provide guidance on which of these capabilities can be obtained outside the Military Service laboratories, which are unavailable except in the Military Service laboratories, and which are so mission-critical (classified, non-proliferation, or other) that they should be maintained in the Military Service laboratories. It will review relevant Department of Defense (DoD) studies and other pertinent literature, and conduct site visits and interviews with all necessary CBDP and DoD Laboratory stakeholders to collect the data to accomplish the study objectives. It will provide guidance for coordination and development of consensus within the CBDP community with a view toward maximizing acceptance and ownership of these requirements and definitions.

Purpose of the Study

This study was requested by the Deputy Assistant Secretary of Defense for Chemical and Biological Defense (DASD(CBD))[1] to assist in several objectives. The committee considered the following objectives as they approached their charge:

1. Defining the terms "core scientific and technology capabilities necessary for conducting core CBDP RDT&E activities" that is acceptable to all DoD CBDP and Laboratory stakeholders. Define how DoD sustains core capabilities, accounting for DoD sustainment guidance, resources, and authorities.
2. Identifying the scientific and technology capabilities DoD requires to accomplish the CBDP mission based on the known and anticipated threats and state of the Science and Technology (S&T) base between now and 2025.
3. Identifying which of these capabilities DoD should consider "core," and recommend core capacity levels at which DoD should seek to sustain core capabilities.
4. Identifying options and needed resources for preserving these capabilities for DASD(CBD) consideration. This includes identifying the current model for sustaining lab infrastructure within DoD and determining if this is the most effective method for maintaining DoD critical infrastructure.

The committee was advised that this report is intended to be used in developing the POM (Program Objective Memorandum) for the Chemical and Biological Defense Program (CBDP).[2]

[1] At press, the current DASD(CBD) is Dr. Gerald Parker.

[2] The POM is used to define funding and programmatic priorities for DoD, and each program must review its own activities and submit that analysis to the agency.

ORGANIZATION OF THE REPORT

This report is organized into five chapters. Chapter 1 describes the statement of task, the committee process, and the purpose for performing the study and provides a brief overview of the amorphous nature of the threat and challenges the CBDP faces. Chapter 2 discusses the CBDP mission and the various frameworks that the CBDP organizational elements use to address their areas of responsibility. The chapter goes on to discuss the committee's view of the mission and introduces the Science and Technology (S&T) Capability Categories that it has identified in order to address the statement of task. Chapter 3 describes core S&T capabilities that are necessary for the CBDP and considers where these capabilities may best be obtained.

In Chapter 4 the committee discusses a strategic capabilities-based planning approach the CBDP could adopt as a way to unify the various program elements around a single set of needs for the program. Chapter 5 discusses the importance of relationships with the end users and the research-development-acquisition (R-D-A) transition, and makes some comments and suggestions for long-term management of the CBDP S&T program, including discussion geared toward the individual DoD laboratories, with the aim of sustaining the core capabilities. Findings and recommendations are found at the end of each chapter and in the Summary.

THE COMMITTEE PROCESS

The committee was asked by DoD to perform this study under stringent time constraints. Over a period of approximately three months, the committee gathered information through a series of briefings, site visits, and interviews with individuals and groups, including stakeholders in the CBDP and operators that rely on the CBDP to support their missions. These included DoD Laboratory personnel; Joint Science and Technology Office for Chemical and Biological Defense (JSTO-CBD), Joint Program Executive Office for Chemical and Biological Defense (JPEO-CBD), and Joint Requirements Office for Chemical, Biological, Radiological, and Nuclear Defense (JRO-CBRND) personnel; CBDP customers, including some US Combatant Commands (COCOMs) and Services; and representatives from organizations and agencies within the broader CBD community, including Cooperative Threat Reduction (CTR) and Global Emerging Infections Surveillance and Response System (GEIS) within DoD, as well as representatives from the Department of Homeland Security (DHS) and the Department of Health and Human Service (HHS).

In addition to the formal briefings and roundtable discussions mentioned above, committee subgroups conducted site visits at US Army Dugway Proving Ground (DPG), US Army Natick Soldier Research,

Development, and Engineering Center (NSRDEC), US Army Medical Research Institute of Infectious Diseases (USAMRIID), US Army Medical Research Institute of Chemical Defense (USAMRICD), and the Edgewood Chemical Biological Center (ECBC). At the various sites, the committee had the opportunity to view relevant facilities, meet with the facility senior leadership, and have discussions with principle investigators and other relevant personnel.

Further details on the committee's data-gathering efforts can be found in Appendix A. The committee also received documents and other materials for review and made specific information requests between meetings that were submitted through the office of the DASD(CBD). Unfortunately, only a fraction of the written information requested was made available to the committee.[3] Despite this, the committee is confident that they were able to fully meet their charge with the information received during briefings and site visits.

Information received during the data-gathering process was examined during committee deliberations and the findings, conclusions, and recommendations are described in this report.

THE THREAT IS AMORPHOUS

The Threat Cannot Be Defined Solely by the Number of Expected Casualties

The United States remains the dominant conventional military force, but experience in a succession of wars—from Vietnam to Afghanistan—have made it clear that a conventional force cannot necessarily respond effectively to non-conventional engagements. The use of improvised explosive devices (IEDs) and explosively formed projectiles (EFPs) provides simple examples: both classes of weapons were well known, but because they were widely available and inexpensive, they have caused politically significant numbers of casualties, and have required disproportionately expensive countermeasures (e.g., armored vehicles, types of operations). The war in Afghanistan is not about exchange ratios or attrition of forces; it is about political advantage, and the patience of the Afghani, Pakistani, and American people regarding the course of the war.

Is there potential for adversaries to create chemical and biological weapons that are as effective as IEDs and suicide bombers? If so, why have they not already been tried? The answer to the first question is

[3] For example, the committee became aware of a report entitled "Chemical and Biological Defense Core Capabilities" released by the Institute of Defense Analyses in March 2007. The committee was not able to review this report during the course of the study.

certainly "Yes"; to the second, the answer is "We don't know." One of the characteristics that have made IEDs so attractive is that explosives of many sorts are readily available during a time of war. A powerful force is familiarity and habit, and if an innovator experimenting with chemical or biological weapons could prove them "successful" (where success would not necessarily be measured in people killed), the technology is readily available for broader use.

Chemical and biological weapons (CBW) have the characteristic that the advantage presently lies strongly with the attacker. There are many possible weapons, from those that are highly developed and very familiar (e.g., hydrogen cyanide, classical nerve agents, anthrax) through more advanced and less well understood threats (e.g., non-traditional agents or NTAs, and tularemia), to hypothetical threats (e.g., genetically engineered viruses) (see Box 1.1). Some of these agents might be used in large-scale, force-on-force engagements; others in "insurgencies of attrition" designed to force the United States to withdraw from a theatre due to financial and political fatigue. It is impossible technically—and unfeasible economically—to try to provide solutions to all threats. There is no analog to "stealth" or "nuclear weapons" or "overhead assets." Scientific and technical innovations that provide such commanding advantage to neutralize the threat for a period of decades do not currently exist. Moreover, because this area of conflict involves everyone at the border between warfare and medicine, including military personnel and civilians on both sides of the conflict, it has—in the United States, but not in hands of some adversaries—regulatory constraints that slow development in ways that are unfamiliar to DoD. The problem is fractal: for every agent or organism, and for every countermeasure, there is a variation—and one that might be easy to implement—that escapes the countermeasure. The United States simply cannot afford to deal with all threats on an individual basis and there is no universal solution. It must choose what problems to solve. Not to choose—the strategy it has largely followed—has resulted in ineffective or uncertain defensive capabilities against many agents. As a result, even in best cases, combatant commanders have a (very) limited ability to estimate the influence of CBW on proposed operations especially when used in innovative ways. (See Appendix B.)

BOX 1.1
Today's Threat to the Force

Historical biological threat

- The primary traditional threat from biological warfare agents was to the force on a distant battlefield in a time of war, by a nation state.
- Until the 1990s, most experts believed a biological attack on the homeland to be very unlikely.
- The biological warfare enterprise within nation states was viewed as similar to nuclear or chemical weapons enterprises.
- The dozen or so agents of concern would have been grown and weaponized on an industrial scale (in ton quantities), at a few facilities at closed sites.
- The predominant threat was thought to be an aerosol.
- Weapons of war—planes and missiles—had been designated to carry the munitions.

Today, the threat to the force worldwide is probably much like the threat to our domestic population

- An attack on the force may be more likely to come from a non-state actor.
- An attack on our US citizens is widely perceived to likely come from a non-state actor or a scientist insider.[a]
- A few grams of biological agent might stop or greatly impact a military operation at home or abroad.
- A few kilograms of biological agent delivered from a motorcycle or boat *could* have near nuclear equivalence.
- While the high-casualty threat is still by aerosol, food (unlikely water) contamination could also cause casualties.
- The small "weapons lab" footprint and the small size of an effective weapon make intelligence very difficult.
- Finally, the United States is not very good at predicting threats of any kind.[b]

The challenges of biological security differ from those of nuclear security[c]

- Much less is known about state biological weapons programs than about state nuclear weapons programs.
- Non-state actors will not have a nuclear program, but might obtain a weapon, or fissile materials that have been produced by others.
- Non-state actors might either obtain a biological weapon or produce one.
- Pathogens are ubiquitous when compared to the few critical isotopes needed for nuclear weapons.
- Few pathogens possess the potential to kill on the scale of a nuclear weapon, but many could cause chaos.
- Pathogens replicate and therefore require different accountability standards than are used for management and control of chemical and nuclear materials

- Biotechnologies are widespread globally in civilian settings versus nuclear technology held by only tens of nuclear states.
- Bioscience, knowledge, and experts number in the tens of thousands rather than in the thousands for nuclear.

Progress in biodefense—against states and non-states—has not kept pace with change in threat

- Sensors and physical protection are still core defenses against the biothreat (for both civilian and military personnel).
    - o The United States probably overestimates their effectiveness in an asymmetric environment.
    - o CONOPS for sensors are not well conceptualized.
    - o Physical protection may only be used after a threat is identified, which would likely occur too late for the protective equipment to be most effective.
- The United States probably has not yet adequately embraced the opacity of the threat.
    - o It will be much, much more difficult to prepare for and defend against than prior threats.
- A robust S&T base to respond to the unknown may be more beneficial than very specific counter measures.

The model for biological security within our service laboratories[d] is similar to nuclear "surety"

1. Personal Reliability
2. Agent Accountability
3 Physical Security
4. Laboratory Safety

There is considerable disagreement across the community regarding the real cost and value of this approach, AR50-1.[e]

---

[a] Graham and Talent; A World At Risk, http://www.absa.org/leg/WorldAtRisk.pdf.
[b] Richard J Danzig; October 26, 2011, Driving in the Dark: Ten Propositions About Prediction and National Security.
[c] Everywhere you look: Select agent pathogens. http://www.upmc-biosecurity.org/website/resources/publications/2011/2011-03-03-select_agent_pathogens.html.
[d] See AR50-1, http://www.fas.org/irp/doddir/army/ar50-1.pdf.
[e] There are four studies from 2008-2009 on this point: NAS, DSB, Trans-Federal Task Force, and NSABB.

There Are Choke Points

What are the choke points in producing technology that would provide effective defense against at least some of the threats the United States and its allies might face? There are a number, not all of them under the control of DoD.

- The threat, although long-standing, is one for which there is very little operational experience, and substantial resistance on the part of DoD to realistic modeling and experimentation. The argument often made is that "there are no good models and simulants." It is probably largely incorrect, but more importantly, when it *is* correct, then *developing* good simulants, to enable realistic simulation, should be recognized as a priority. Without realistic information about the problem, one cannot develop a solution.
- There is almost no red-teaming—a critical weakness when there is no operational experience. There is no prospective examination of how chemical and biological weapons might be used in innovative ways in the hands of adversaries with different legal, ethical, and financial constraints than those under which we operate, and with different operational and political goals and objectives. The approach used would never anticipate, for example, the chemical and biological equivalent of a suicide bomber, or an EFP, or a building denial weapon. At least some of the responsible laboratories seem to not be committed to identifying problems and finding solutions, but rather to continuing existing, and often outmoded, themes.
- The rules constraining acquisition within DoD, regulatory clearance, clinical trials, and negotiations with the US Food and Drug Administration seriously slow the development of vaccines and therapeutics intended for *in vivo* use in humans, as well as the adoption of modern highly multiplexed diagnostics.
- US universities—the best in the world in almost all areas of applied biology, genomics, biochemical mechanisms of disease and toxicity, biomedical materials, and many other relevant areas—are widely unengaged in problems specific to DoD needs.
- The US pharmaceutical industry also has not been effectively engaged, in part because of strict financial rules that come with working with the government, and in part because military markets are too small to be interesting financially. DoD has been ineffective in formulating programs that would engage large pharmaceutical companies. The problem is made more complex by the fact that the pharmaceutical industry is one of the most

internationalized, and there are few big pharmaceutical companies that only operate in the United States.
- The mindset underlying the prioritization and evaluation of programs has *not* been informed and skeptical, as a good engineering program should be. Rather, it has often pursued technology based on unrealistic objectives and evaluated against weak and unrealistic standards with little or no consideration of cost or practicality.

New Tools Are Available

To balance the difficulties of this area, however, there is also an explosive growth of relevant biological and biomedical science, deep interests shared by military and public health sectors, and a biomedical establishment that is the best in the world.

The problem of CBD has been in the category of "too hard" for decades but new breakthroughs hold great promise for this problem:

- The reduction in the cost of genomic sequencing has been remarkable.
- The pharmaceutical industry has developed an extraordinary competence in toxicology, mechanisms of infectious disease, and resistance of pathogen-based disease to therapeutic agents.
- Analytical instrumentation for bioanalysis (as applied in biomedical research and civilian medicine) has developed very rapidly.
- Industrial toxicology—driven both by regulatory requirements and by environmental concerns—has begun to move from pure engineering to engineering supported by science.
- Reporting of disease and data-based public health has become a rapidly evolving area.
- Simulation and modeling has developed enormously as computers and user interfaces have developed.
- Drones and robots have developed rapidly and have the potential to augment existing sensors and collectors for some tasks.

Relatively little of this technology has been transferred to organizations responsible for CBD, so, for appropriate problems, there is the potential for significantly increasing capabilities just by using existing technology. The DoD effort in CBD has also been hindered by its inability to form close connections with academic laboratories, and by the practice of keeping much of the research internal to DoD laboratories. In fact, almost none of the fundamental science required for CBD lies primarily in DoD, although specific applications of that science to threat agents likely do.

Metrics for Success

In the absence of operational experience against a determined and innovative adversary using live weapons and counting real casualties, how is the United States to understand whether its technology, objectives, and strategy are effective? It seems clear that the only replacement for reality is realistic experiments (red team on blue team) using realistic simulants, and computer modeling.

The computer modeling in this area is a crutch: it is safe and inexpensive relative to real exercises, but ineffective in providing solid engineering information to use in building better and more realistic models. The lack of empirical data on which to base the models is an important contributor to the absence of useful models. The experimentation carried out by DoD in the CBDP is strongly academic, and usually so removed from plausible use that it is difficult to believe that it would help the COCOMs to understand and evaluate the program's impacts, how best to protect their forces, to carry out their operations in the face of these weapons. They would figure it out with time (as, with the help of other countries, they figured out the IED problem), but the time required to do so could have profound impact on operational readiness and effectiveness and on the course of missions.

PROPOSED APPROACH

The foundation of addressing the complexity of chemical and biological defense is to begin with clarity of purpose in what DoD aspires to achieve. This means describing

1. the scope of events that could unfold to threaten national security,
2. the range of strategies for responding to these events,
3. how the consequences of events and performance of the response are measured, and
4. the choices made about which defense and response capabilities to implement and grow, necessarily reflecting tradeoffs among capabilities, risks, and costs.

Describing each of these elements of a chemical and biological defense strategy is difficult. Planning tends to focus on narrow conceptions of threats and responses derived from historical events. Outcomes tend to be described in terms of consequences which can be easily measured, such as fatalities and injuries. Options tend to be developed based on incremental modifications to current materiel and operations. Each of these approaches is inadequate for addressing the innovative nature chemical and biological threats.

Instead, planning must expand the range of options considered in each element. Iterative review and realistic red-teaming challenge assumptions built into plans and promote innovations in defense to successfully respond to the threats. The scope of red-teaming and review should encompass the threats and activities against which performance is assessed and the evaluations of performance are made.

Doing this type of planning is difficult and requires dedication of resources and consideration of time to allow it to occur. There are numerous examples of attempts to do this poorly. Chapter 4 provides more details on how to implement this approach to strategic planning. When used, it leads to informed choices about implementable strategies to improve security and defense capabilities.

FINDINGS AND RECOMMENDATIONS

In exploring the state of the chemical and biological warfare threat, the committee identified two principle findings:

Mission and Strategy

Finding 1.1: The threat is unpredictable, changing, and dependent on the nature of conflict. The CBDP cannot rely on breakthroughs in intelligence on adversaries' chemical or biological terrorism or warfare programs to inform how its investments are prioritized.

Finding 1.2: The program has not adapted to the changing nature of the chemical and biological threat. It is impossible technically—and unfeasible economically—to try to provide solutions to all potential threats. The United States simply cannot afford to deal with all threats on an individual basis, and there is no universal solution—it has to choose which problems to solve.

2

Framework and Structure

ORGANIZATIONAL CONCERNS

The Chemical and Biological Mission Is Broad and the Strategy Is Unclear

CBDP Mission: "Provide global chemical, biological, radiological, and nuclear defense capabilities in support of National Strategies"

A contribution to the persistent problem space, besides the organization, is that the Chemical and Biological Defense Program (CBDP) mission statement is overly broad, allowing wide-ranging interpretation of what is included and, therefore, what is most important. Among the four principle missions are

- Warfighting: Operate Through,
- Homeland Defense,
- Warfighting Support Functions, and
- Consequence Management (Foreign or Domestic).

Among these four missions, priorities vary within each office and with leadership changes each shift priorities in terms of resource allocation and programmatic emphasis.

In the last decade and a half, the CBDP has seen its strategy shift—appropriately in the view of the committee—as leadership has sought a better balance between chemical threats (previously dominant) and biological threats (of greater concern post 2001 anthrax mailings), and to address the rise of sub- or transnational perpetrators. Accompanying the strategy shift have been changes in investment priorities, but not necessarily in a comparably balanced way. Seeking to have impact quickly, leadership first de-emphasized contamination avoidance in favor of broad-spectrum medical countermeasures. More recently that too has shifted to greater international cooperative public health monitoring and threat reduction. The changes have resulted in rapid ramp-up of investments initially in the Transformational Medical Technologies Initiative (TMTI) in 2005, followed more recently by a decline in TMT funding[1] but upticks in biosurveillance and the biological aspects of the Cooperative Threat Reduction program.

While it is difficult to argue against the importance of any of these principle mission areas, all of them demand sustained investments over many years to achieve both fielded capabilities and a robust S&T base to support and advance the state of the art against threats that are not static. Significant variations in funding over three- to four-year cycles are unlikely to yield much progress. For example, the complexities of the development and approval process for a new medical countermeasure require a commitment of approximately eight to ten years. A different set of challenges faces implementation of an international biosurveillance network as it will require that information from many, varied data streams must be carefully integrated in order for analysis to be meaningful.

The committee believes in the wisdom of a defense-in-depth strategy, which balances investments end to end, from pre- to post-event elements. Moreover, sustained investments over appropriate periods of time are required. This approach does not translate to guaranteed funding to performers—they should still be accountable for progress—but it does mean that program decisions come with leadership commitment of continued support as milestones are met on the path to fielded new or improved capabilities.

[1] The committee heard from several former TMT performers; one possible conclusion is that TMT became a "Big Science" project that outgrew the CBD organizational structure and management culture that was geared toward managing small, independent projects within distinct divisions (e.g., Diagnostics, Medical Countermeasures, etc.). It appears likely that this small-science managerial culture still predominates throughout the Department of Defense (DoD), which has potential implications for the newly growing programs, particularly since their success requires effective managerial coordination across both the JSTO-CBD and the JPEO-CBD.

Organization of the Chemical and Biological Defense Program

History

Prior to the first Gulf War (1991), chemical and biological defense science and techology (S&T) was conducted in the individual Military Services, with the Army being the primary participant (supplemented by limited interest, funding, and execution in the other Services). While the threat was generally known throughout DoD, and perhaps the general public, until the confrontation with Iraq there had never been an imminent chemical and biological (CB) threat. Fortunately, there was no documented use of either a chemical or biological agent by Iraq against US forces. Nonetheless, the perceived lack of adequate protection against chemical and biological agents rose to an unprecedented level of interest, both within DoD and Congress.

Congressional interest resulted in the enactment of PL 103-160 in 1993. This statute consolidated the CBDP into a joint program, with "oversight" at the Office of the Secretary of Defense (OSD). To address the need for Joint Service participation and execution of the program, two groups were established. First, the Joint Service Integration Group (JSIG) was directed to identify Joint requirements and to lead development of a Program Objective Memorandum (POM) that would address the needs of the Services. Second, the Joint Service Materiel Group (JSMG) was created to oversee the research, development, test, and evaluation (RDT&E) functions of the CBDP. The JSMG created the Joint Program Office (JPO) to manage the advanced development aspects of the program. The Army took the lead for creating this office and it essentially reported to the Assistant Secretary of the Army for Research, Development, and Acquisition (ASARDA), thus creating a bifurcated reporting chain. The JPO was originally headed by an Army Colonel (Chemical Corps background), with a civilian deputy. Subsequently, the Joint Program Executive Office for Chemical and Biological Defense (JPEO-CBD), successor to the JPO, was recognized as an Acquisition billet and therefore filled with appropriately trained individuals, while maintaining a senior civilian as the deputy. Both the JSIG and JSMG had flag officer members, and theoretically reported to the OSD to comply with the mandate that OSD be the single focal point within DoD. This somewhat cumbersome management structure remained in place for approximately 10 years. Changes in leadership at various levels resulted in changes in program direction and varying levels of interest, oversight, and advocacy for the program from the OSD.

The second Gulf War in the early 2000's led to the difficult realization that not much had changed in the CBDP in the preceding decade. Subsequently, a memo from the Under Secretary of Defense for Acquisition, Technology, and Logistics, issued in 2003, created the current program management structure. In this arrangement, management of the S&T programs was assigned to the JSTO-CBD within the Defense Threat Reduction Agency (DTRA), the advanced development programs continued to be managed by the JPEO-CBD, and the responsibility for establishing the requirements for the CBDP was assigned to a new office within the J-8 of the Joint Staff (the Joint Requirements Office for Chemical, Biological, Radiological, and Nuclear Defense (JRO-CBRND)). This system, represented in Figure 2.1, has created a program management structure which has different reporting chains (i.e., the JSTO-CBD reports to DTRA leadership; the JPEO-CBD reports to ASARDA; and the JRO-CBRND reports to the Joint Staff). Oversight at the OSD level is limited, with the Deputy Assistant Secretary of Defense for Chemical and Biological Defense (DASD(CBD)) having no mandated authority to change program direction, influence the POM and budget, or coordinate individual program/project efforts. This management structure, not surprisingly, is far from effective. Despite claims to the contrary, the committee observed that individual components work independently, with no clear definition of mission and no clear allegiance to program leadership.

The Office of the Assistant Secretary of Defense for Nuclear, Chemical, and Biological Defense Programs/Chemical and Biological Defense (OASD(NCB/CB)), JRO-CBRND, JSTO-CBD, and JPEO-CBD each view the chemical and biological defense mission from a different perspective. Although this is appropriate to some degree, the coordination and collaboration between these groups is far from seamless, which limits their ability to contribute to the mission and vision of the program as a whole.[2]

- JRO-CBRND views the program in terms of operational elements (i.e., Sense, Shape, Shield, Sustain) and functionalities (e.g., chemical point detection and biological prophylaxis).
- JPEO-CBD views the program in line with the JRO-CBRND operational elements in terms of acquisition and fielding of products (e.g., Joint Chemical Agent Detector, anthrax vaccine).
- JSTO-CBD views the program in terms of S&T capabilities (four thrusts and seven enablers), with the focus on research and early development.

These different views are summarized in Figure 2.2. Overall, there is a lack of clarity within CBDP regarding the full scope of the mission space. Though all parts of the program agree on the dominance of the warfighting mission, there is not necessarily agreement on whether the goal of CBD should be enabling warfighters to "operate through" or if

[2] See Appendix C for individual framework descriptions.

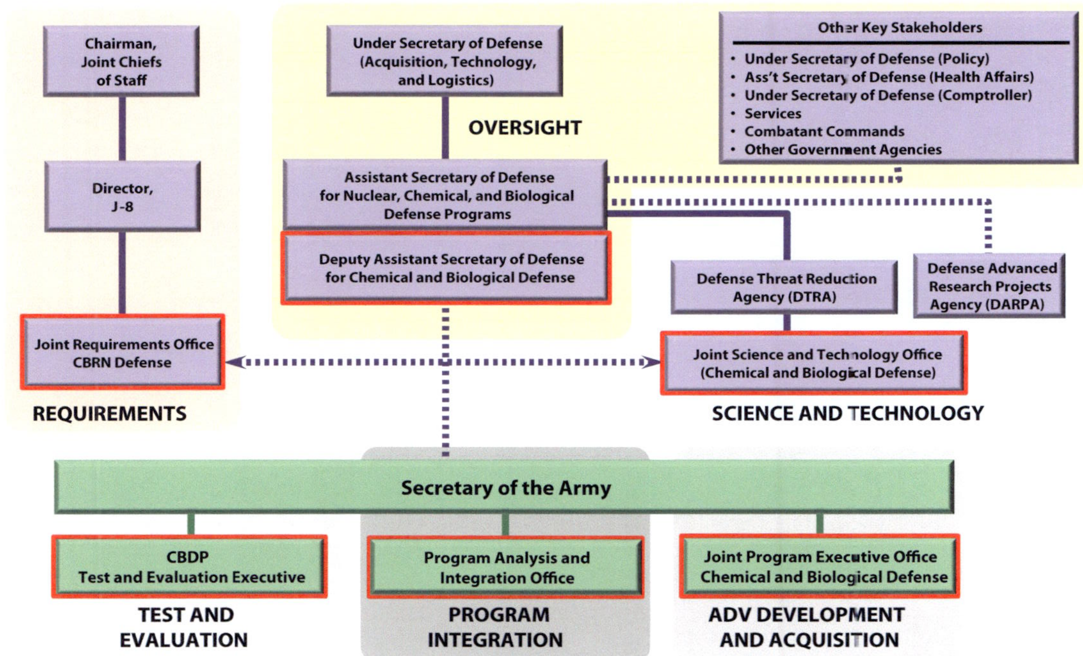

FIGURE 2.1 Organizational chart of the CBDP enterprise. Solid lines indicate formal line of authority; dotted lines indicate necessary coordination, but no formal lines of authority. The principal program players in the CBDP are highlighted with red borders. The dotted line or no line connectivity among them is factual. No hard-line relationships exist among these offices and the committee observed that indeed each appears to act on its own, with separate organizing constructs as portrayed in the framework comparison of Figure 2.2. There is, in fact, no program-wide chemical and biological defense strategy, or common characterization of the program elements.

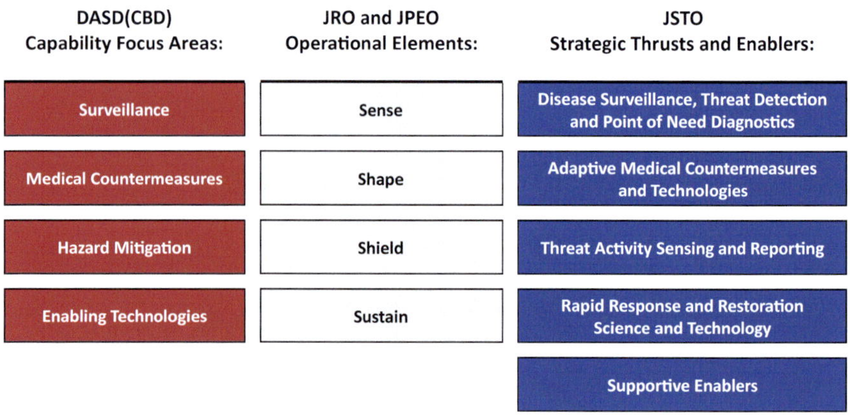

FIGURE 2.2 Chart comparing the high-level frameworks, primary CBDP enterprise elements.

it is achieving zero casualties after exposure to CB threats. In addition, these different missions mean that many parts of the organizations do not share a common understanding of what priority should be placed on the support missions (either for the warfighter or to civil authorities). All of this is concerning for both CBDP's development and acquisition functions because ambiguity about mission priorities could lead to overspecified, underspecified, and/or unspecified requirements, or worse, no fielded capability.

THE COMMITTEE'S APPROACH

Taking into consideration the multiple perspectives of the CBDP offices, the committee found that it needed both common definitions and a common framework in order to determine the core S&T capabilities that are required for the CBDP.

The committee agreed on a series of definitions to discuss the CBDP capabilities. The chemical and biological defense enterprise is encouraged to maintain a well-defined set of terms for use in the CBDP enterprise. The working definitions for this report are:

- **CBD operational capability:** Ability to assure success of the military mission by avoiding or operating in/through a chemically or biologically contaminated environment.

- **S&T capability:** RDT&E capacity and RDT&E competency to enable development, improvement, and/or innovation of a CBD operational capability.
  a. **RDT&E capacity:** Resources (e.g., equipment, facilities, people, plans, funds, etc.) that enable a capability.
  b. **RDT&E competency:** Quality at which a function or service can be provided, as recognized by outside experts.
- **Core (adj.):** Essential to ensuring the capability.

Using these definitions, the committee began to think about a framework for the development and discussion of Core S&T Capabilities for the CBDP. Taking advantage of the Joint Requirement Office's Chemical, Biological, Radiological, and Nuclear (CBRN) Defense Operational Elements[3] (Shape, Sense, Shield, Sustain) and the CBDP Strategic Framework Constructs from the office of the DASD(CBD)[4] (Prevent, Protect, Mitigate, Respond, Recover), the committee mapped these elements to the four overarching missions as described in Figure 2.3.

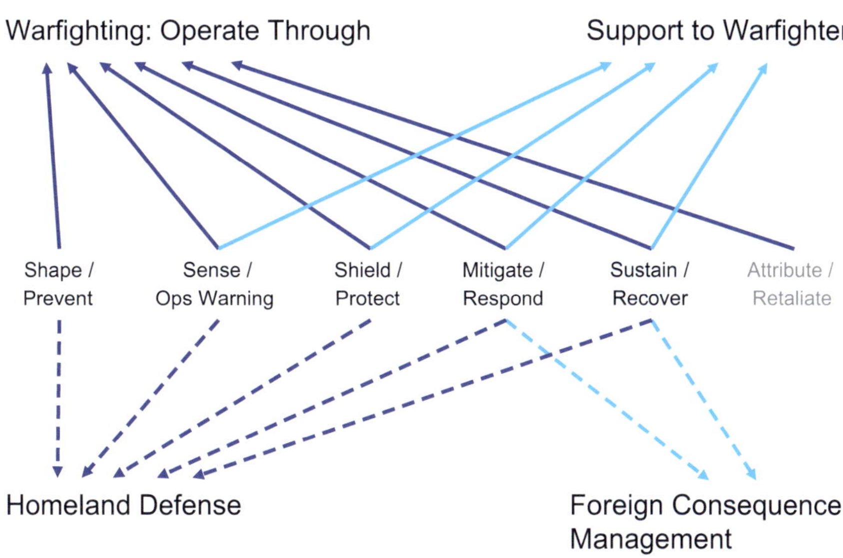

FIGURE 2.3 Overview chart of the relationship between the four mission areas of CBDP and the program's six operational elements.

[3] Modernization Plan for Chemical, Biological, Radiological and Nuclear (CBRN) Defense, Joint Requirements Office for CBRN Defense, January 5, 2012.

[4] Chemical and Biological Defense Program (CBDP) Strategic Framework and Department of Defense (DoD) Organization, Dr. Robert Cohn, Chief Scientist, Office of the Assistant Secretary of Defense for Nuclear, Chemical and Biological Defense Programs/Chemical and Biological Defense, February 15, 2012.

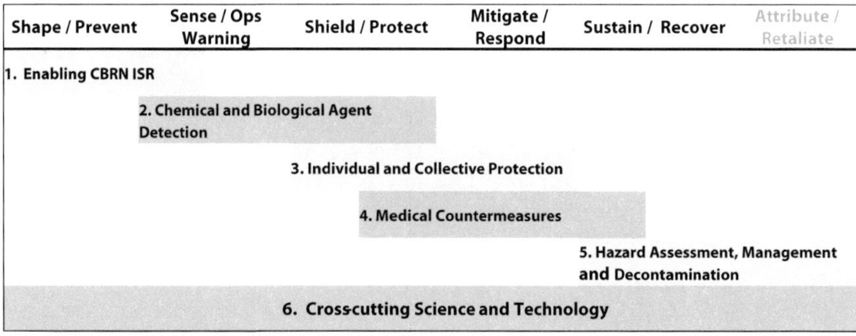

FIGURE 2.4 Chart depicting the relationship between the six CBDP Operational Elements and the six CBD S&T capability categories.

Once the appropriate correlations between the CBD Operational Elements and the missions of the CBD program were identified, they were used by the committee to identify S&T capability categories required to address both the missions and the defined operational elements. The six committee-identified CBD S&T capability categories can be discussed based on their mapping to the "time oriented" operational elements. A listing and mapping of the CBD S&T capability categories is found in Figure 2.4.

The CBDP is of such broad scope that it has a role to play in all stages of chemical and biological defense. While the CBDP is not a direct actor in intelligence, surveillance, and reconnaissance (ISR), the program plays a role in the development of ISR S&T tools. To this end, the committee identified (1) **Enabling CBRN Intelligence, Surveillance, and Reconnaissance** as one of the core S&T capability categories. The program plays a greater role in (2) **Chemical and Biological Agent Detection** which supports Sense/Operational Warning and Shield/Protect operational elements. The operational element Shield/Protect, coupled with Mitigate/Respond supports the (3) **Individual and Collective Protection** S&T capability. (4) **Medical Countermeasures** falls under Shield/Protect, Mitigate/Respond, and Sustain/Recover, while (5) **Hazard Assessment, Management, and Decontamination** is supported by overlapping Sustain/Recover with Attribute/Retaliate, an operational element that is only on the edges of the purview of the CBDP. Finally the committee identified several S&T capabilities that are core to all of the identified operational elements. These capabilities are covered in (6) **Cross-Cutting Science and Technology**.

As stated previously, the committee found many operational frameworks present within the units of the CBDP enterprise. Each unit views the problem through its own lens. With the framework for viewing the core S&T capabilities in hand, the disparate frameworks of OASD(NCB/CB),

JRO-CBRND, JSTO-CBD, and JPEO-CBD fit under the six S&T capability categories defined by the committee (see Appendix D for a chart comparing the S&T capability frameworks of the CBDP enterprise elements in relationship to the committee's six CBD S&T capability categories).

The next chapter will expand on the S&T capability categories, identify the core S&T capabilities that are need to be maintained by the CBDP, and consider where the capabilities may be obtained.

FINDINGS AND RECOMMENDATIONS

In Chapter 2 the committee identified the following findings and recommendations:

Mission and Strategy

Finding 2.1: The CBDP mission is too broadly stated. The stated mission of CBDP is to "Provide global chemical, biological, radiological, and nuclear defense capabilities in support of National Strategies." The mission statement is large enough to allow for a wide variety of interpretations, making it challenging for both the customers of the program and the facilities that support its work to understand the program priorities.

The CBDP has responsibilities that span missions from protecting the warfighter and providing support to the warfighter, to defending the United States from attack (i.e., Homeland Defense) and supporting local authorities following a chemically or biologically related incident (i.e., Consequence Management, Foreign or Domestic). Events requiring the Department to perform each of these missions could unfold in innumerable, unexpected ways; for example,

- naturally occurring disease or unintentional chemical exposures may be difficult to distinguish from intentional attacks;
- intelligence, as noted above, has historically proved uncertain and/or unreliable in assessing the CB threat;
- innovations, as witnessed in the case of improvised explosive devices, will certainly occur in the development and use of chemical and biological weapons; and
- the United States has a poor understanding of the intentions of those who might use chemical and biological weapons, and an even poorer understanding of the barriers that prevent them from doing so.

The combination of a broadly stated mission and numerous uncertainties calls for top-level guidance on focus and priorities, which the current program lacks.

Finding 2.2: There is no program-wide CB defense strategy, nor common characterization of the program elements among the participating organizations. The many different organizations currently involved in the CBDP (e.g., OASD(NCB/CB), JRO-CBRND, JSTO-CBD, and JPEO-CBD) each view the chemical and biological defense mission from different perspectives. Although this can be expected based on their different roles, the coordination and collaboration between these groups is far from seamless. As a result, the program has been—and continues to be—limited in its ability to deliver fielded solutions in a timely manner.

Finding 2.3: Strategic priorities tend to change with changes in senior leadership. As a result, efforts requiring sustained and/or longer-term commitments (e.g., medical countermeasures) are unable to deliver timely results, if at all.

Recommendation 2.1: The DASD(CBD) should lead a mission and strategy development activity that aligns all of the program elements and offices. The differences among offices in how they portray and communicate their stories, in their priorities, and in the terminology they use to describe the program are stark. Bold moves are needed to break the current stagnation that permeates the chemical and biological S&T and acquisition environment. Tweaking the management or refocusing a few projects will not be sufficient. The recommended alignment activity should promote a shared understanding of and commitment to key priorities for maintaining the core capabilities and expertise needed to fulfill the overall program mission and strategy.

3

Core Science and Technology Capabilities for the Chemical and Biological Defense Program

The committee explored the various science and technology capabilities that are relevant to the Chemical and Biological Defense Program (CBDP). In order to address its task, the committee utilized the six science and technology (S&T) capability categories defined in the previous chapter: **Enabling CBRN Intelligence, Surveillance, and Reconnaissance; Chemical and Biological Agent Detection; Individual and Collective Protection; Medical Countermeasures; Hazard Assessment, Management, and Decontamination; and Cross-Cutting Science and Technology.** In addressing the first five areas, excluding Cross-Cutting Science and Technology, the committee identified four to eight S&T capabilities for each category that "must be available to support Chemical and Biological Defense (CBD) research, development, test and evaluation, and operational activities."[1] Table 3.1 shows the S&T capabilities identified by the committee under the six S&T capability categories.

Except where otherwise noted, each of the identified S&T capabilities is of sufficient importance to the CBDP that it should provide support to further the activity. To that end, this chapter describes the committee's views on which capabilities can and should be obtained in Department of Defense (DoD) laboratories and which are better suited to be obtained outside of the DoD or outside of the government altogether. To reach these conclusions, a decision tree was developed with several elements

[1] Statement of Task.

TABLE 3.1 Committee Identified S&T Capabilities

1. ENABLING CBRN INTELLIGENCE, SURVEILLANCE, AND RECONNAISSANCE	2. CHEMICAL AND BIOLOGICAL AGENT DETECTION	3. INDIVIDUAL AND COLLECTIVE PROTECTION
Information Acquisition and Analysis	Analytical Methods Discovery	Controlled Molecular Transport Materials Discovery
Health Monitoring	Instrumentation Development	Barrier Materials Engineering
Environmental Monitoring	Sensor Systems Development	Personal Protective Systems Development
Unknown Agent Identification and Characterization	Agent Transport Analysis	Collective Protection Systems Development
		Physiology
4. MEDICAL COUNTERMEASURES	**5. HAZARD ASSESSMENT, MANAGEMENT, AND DECONTAMINATION**	**6. CROSS-CUTTING SCIENCE AND TECHNOLOGY**
Target Discovery	Decontamination Methods Discovery	Acquisition, Maintenance and Transport of Critical Chemical and Biological Reagents
Regulatory Science	Decontaminant Development	Agent Simulation
Mechanisms of Delivery and Delivery Systems	Decontamination Resilient Materials Development	Informatics
Animal Models	Decontamination Systems Engineering	Statistical Measurement Design
Host Response	Agent Transport and Viability Analysis	Forensics
Pre-Clinical Studies		Education and Training
Clinical Trials (GLP and GMP)		Behavioral Analysis
Medical Product Development		Systems Analysis and Engineering
		Repurposing Commercial Technologies
		Systems Biology
		Synthetic Chemistry and Biology
		Materials Science
		Test and Evaluation

to consider at each node in order to help address the necessity of DoD to maintaining certain capabilities in-house (see Figure 3.1).

For the S&T capabilities listed, which the committee has already determined to be core to the CBDP, the committee asked the following questions:

(1) Should the CBDP be providing funding to support this core S&T capability?

The obvious answer may well be yes, but the committee explored some possible reasons that the CBDP should avoid using its limited resources in a specific area. For example, if another part of DoD or the government, which is viewed as reliable and competent, is already funding this area, maybe the CBDP should not invest heavily in this area, provided the two government units involved communicate with one another and make shared use of the findings.

(2) Assuming the CBDP does decide to support and fund a specific S&T capability, the committee looked at whether this capability *can* be found outside of the government.

As defined previously, capability is having both competence and capacity to accomplish a goal. Do the extra-governmental sources have the competence and capacity, the ability to ensure safety, security and scale to a given problem? If not, should DoD, as opposed to another governmental agency, be focusing on the S&T capability (see step 4)?

(3) If it *can* be found outside, *should* the government do it anyway?

In addressing this question, the committee considered a variety of factors that are of high importance to the program. What is the cost differential? Is there both long- and short-term domestic availability[2] in the private sector? Does industry provide the necessary staying power for CBDP? Are there safety and security considerations that would require the program to maintain control? If these issues can be satisfactorily addressed, then perhaps CBDP should obtain this capability primarily from the private sector.

(4) If the above questions direct the government to maintain an internal capability, the CBDP should then consider whether it should maintain the capability itself, within DoD, or obtain it from other

[2] The meaning of availability will be different for each S&T capability. Considerations for intellectual property rights are important in making this determination. Procedural barriers, such as contracting, should be considered in the short term, but care must be taken not to ascribe all problems to "procedural barriers" to avoid finding a solution.

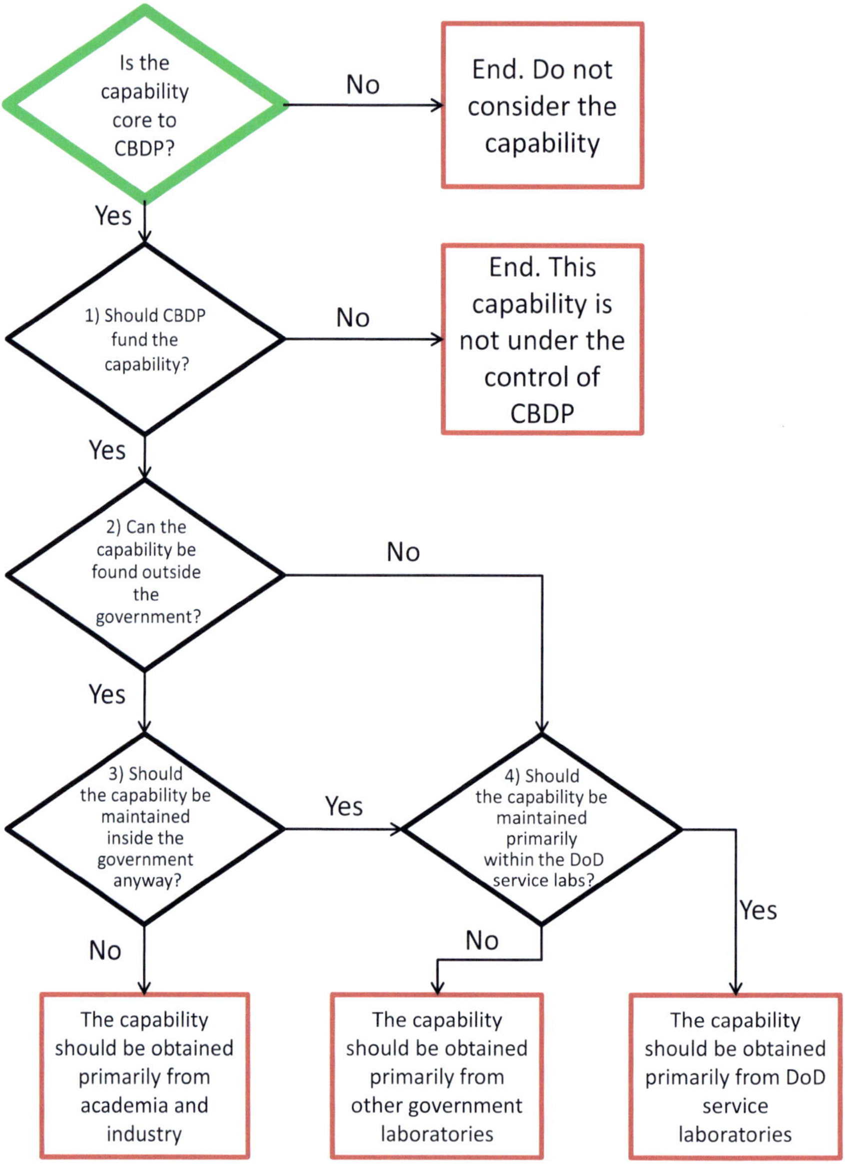

FIGURE 3.1 The decision tree for determining if capabilities should be housed within DoD or leveraged from external sources.

government laboratories[3] (e.g., Centers for Disease Control and Prevention, National Institutes of Health, Department of Energy National Laboratories). Items to consider when making this assessment include agency mission, agency resources, and possible public health overlap.

While the committee does not necessarily believe that it should be a high bar for the CBDP to maintain in-house, DoD-controlled capabilities, there are some important questions to consider when discussing the location of the various S&T capabilities.

The committee used the decision tree (Figure 3.1) to determine whether individual capabilities should be maintained by the program and where. Many of the metrics used to address each decision node on the tree are subjective and the committee's consensus view is described in this chapter. If the CBDP does not agree with an individual assessment of an S&T capability, they are encouraged to undertake a *de novo* analysis of that capability, using the decision tree above,[4] to reach their own conclusion.

ENABLING CBRN INTELLIGENCE, SURVEILLANCE, AND RECONNAISSANCE

The CBDP must have a natural and fundamental role in the prevention of and strategic warning against threats to global and national security through the use of or exposure to priority biological and chemical agents. No other program is better positioned to project capability forward or contribute to global strategic, operational, and tactical warning and prevention. From the force health protection perspective, CBDP plans and provides resources for the research, development, test and evaluation (RDT&E) and delivery of vaccine, prophylaxis, and therapeutic technologies that protect US military personnel and civilians, as required when exposure is determined or suspected. Further, the CBDP supports and directs funding for detection, diagnosis, and biosurveillance, which simultaneously aids in advancing forensic capabilities that inform deci-

[3] We make a distinction between non-DoD government labs and non-government performers. This is based on an assumption that government labs meet security and surety standards (e.g., ability to work with classified samples, materials, and information) and are designed to be an enduring capability that will be available when needed. Non-government labs have no such assumption of guaranteed continuity, even if security and surety standards are currently met.

[4] The real intent of the committee's approach is to illustrate how to apply a structured, systematic, and consistent process to decision making that allows for a common understanding of priorities and how they were developed. The results also provide a basis for continuity of support.

sions related to medical protection and treatment as well as those that inform intelligence, operations planning, medical intelligence, and attribution. Just as S&T contributes considerably to medical decisions, it also contributes to military and policy decisions related to possibly nefarious activities involving biological and chemical agents. The committee has identified four S&T capabilities that are required to provide technologies needed for chemical, biological, radiological, and nuclear (CBRN) intelligence, surveillance, and reconnaissance (CBRN ISR): information acquisition and analysis, health monitoring, environmental monitoring, and unknown agent identification and characterization.

Information Acquisition and Analysis

The core concept for this capability is that multiple dense sources of information pertinent to CBRN ISR exist and need to be properly mined, managed, integrated, distilled, and efficiently queried. A few examples of information sources in the biological domain include ProMed Mail (outbreak reports), ARGUS (a system that searches World Wide Web news media for signs of social unrest that could be due to pathogen outbreaks), Global Public Health Information Network, social media such as Facebook or Twitter,[5] and global weather forecasts and reports. To be fully effective, an information acquisition and analysis system needs to be capable of extracting pertinent facts from a large number of languages and information source types. This involves numerous difficult and largely unsolved problems in natural language processing and information fusion.

DoD needs to be well informed of potential and actual biological and chemical outbreaks and incidents as they relate to both existing and potentially needed detection capabilities. There are multiple government and non-government sources of relevant information for enabling CBRN ISR. Additionally, there are multiple government-funded groups that already attempt various levels of information integration in the chemical and biological domain. It is likely that no existing systems individually meet all current DoD needs. However, DoD should strive to leverage as much as possible from collaborations with existing systems, rather than launching any *de novo* efforts at developing such capabilities. The committee noted that the current program plans to search social media for outbreak-relevant information did not appear to have a plan for integrating with other available information sources. CBDP should be better aligned with other relevant government information acquisition and analysis efforts in the biological and chemical domains.

---

[5] We note that legal issues involving the Privacy Act are faced by any information systems that mine information about US citizens, even if such information is in the public domain.

Health Monitoring

US forces and civilian personnel are projected and mobilized in and out of a diverse array of operational settings and environments. The DoD must be aware of, and be able to protect and respond to, threats to health and readiness. Not only should the DoD support protection and prevention against routine diseases common within the continental United States,[6] it must also be able to provide agile medical and S&T responses to "surprise" in any location worldwide. Thus, programs to address those diseases that are intentionally caused, endemic or presenting elsewhere in the world where DoD operates, need to be incorporated into CBDP planning, resourcing, and delivery. While every new disease cannot be anticipated, a system and capabilities must be in place for CBDP to respond effectively and in a timely manner. This applies to medical treatment approaches as well as diagnostics, biomarkers to exposure, and biosurveillance.

One aspect of this is to consider health monitoring in the context of "One Health,"[7] which considers human health in the global context of animals (domestic and wild) and includes all reservoirs and vectors (e.g., mammals, birds, ticks, mosquitoes, etc.). This implies increased global biosurveillance of both human and animal baselines to determine what is currently endemic so that new pathogens with pandemic potential can be more rapidly detected and characterized, ideally before they are spread widely via global air transportation. Effective health monitoring implies the development and use of a broad hierarchy of diagnostics that includes not only inexpensive point-of-care presumptive diagnostic kits (akin to home pregnancy tests), but also sensitive and precise multiplex confirmatory diagnostics (symptomatic panels such as respiratory, fever, enteric, encephalitis, etc.; also panels for food-borne pathogens, zoonotic pathogens, etc.), broad-spectrum microarray diagnostics (as a safety net to cover unusual/unexpected known single pathogens or combinations), and finally, sequencing to detect and begin to characterize unknown pathogens. It is important to note that global biosurveillance cannot consist of merely the high-consequence pathogens that the United States is most concerned with. It should offer real benefit to both the local communities and the US warfighters who may be serving there, in terms of diagnosing endemic pathogens of local concern. To be effective, the

[6] Traditionally, force protection against natural health threats was primarily addressed by WRAIR, and force protection against biological weapons was the main focus of USAMRIID; recently, however, biothreat agent funding has been approved for use in addressing pandemic threats, such as H1N1. The committee recognizes that this decision has broad implications, but believes that an examination of this issue is beyond the scope of this report.

[7] Also called "species-neutral."

United States cannot expect to merely "cherry-pick" samples of potential pandemic or bioterror cases.

Another aspect of health monitoring is that it requires the ability to be adaptive in order to respond to unknown pathogens as they appear. There may be a truly novel heretofore undiscovered/unrecognized pathogen, or bio-engineered pathogen. New "signals" must be identified and the proper responses provided for. "Signals in noise" must be discerned and put in context with and measured against routine disease outbreaks. A robust system of integrated biosurveillance, through the proper "kits" of low-, medium-, and high-resolution diagnostics, best position the CBDP to provide agile and comprehensive support to the warfighter. This is true whether the focus required is force health protection, humanitarian assistance, global biosurveillance, or intelligence to support operational or policy decisions.

Despite the fact that other government agencies have been involved with pathogen detection for many years, the CBDP does not currently have the required completed hierarchy of Food and Drug Administration (FDA)-approved diagnostics outlined above. Nor does it appear that there is any defined path through other government agencies to achieve this goal, for reasons that are beyond the scope of this committee.[8] Thus, it appears to be imperative for CBDP to take up the charge of ensuring that the warfighters are provided optimal protection by ensuring that appropriate FDA-approved diagnostics are available to global biosurveillance efforts, as well as available for use in DoD medical facilities in theaters of operation worldwide. To the extent that US regulatory policies and procedures impede the timely delivery of modern diagnostics and effective countermeasures to the warfighter, CBDP should continue to actively engage with the FDA to provide needed improvements.

The responsibility to ensure that the DoD health monitoring needs[9] are achieved clearly lies within the DoD. Ideally, more diagnostics could be leveraged from the Department of Health and Human Services (HHS) via the National Institutes of Health (NIH) and the Centers for Disease Control and Prevention (CDC), and/or the United States Department of Agriculture (USDA) for animal or other host/vector diagnostics, but it does not appear likely that timely completion of the required hierarchy of

[8] As discussed in Chapter 5, research-development-acquisition of medical products face unique challenges, including obtaining FDA approval. FDA-approved diagnostics may be an unrealistic goal, especially for diseases endemic to developing countries, as funding to support the development of FDA-approved diagnostics is not supported by the eventual potential market for those tests.

[9] DoD health monitoring needs for the warfighter and civilian personnel are to be aware of, and be able to protect and respond to, threats to health and readiness, including those diseases that are endemic or presenting wherever DoD operates.

diagnostics will be achieved via these routes.[10] Much of the work to create and test pathogen diagnostics could be performed outside of DoD and other government agencies (under appropriate government contracting methods). However, live-agent testing for many high-consequence pathogens may need to continue to be performed within government facilities to ensure the ability to do so.

Environmental Monitoring

The DoD desires to operate in a "detect to warn" mode, which translates most broadly into "detect an approaching cloud of chemical or biological agent swiftly enough to issue a warning for the warfighters to don their protective gear." Thus, the committee's operational definition of environmental monitoring is what is known as "standoff detection," and typically would be utilized in the event of either a deliberate attack (chemical or biological) or an industrial accident (most likely chemical, but potentially biological).

Standoff detection is a perennial high-priority request for CBDP for base protection. Standoff detection of potentially dangerous chemical or biological materials is a difficult goal to achieve considering the relationship of distance and agent concentration upon signal detection. Achieving acceptable false positive and negative rates, considering the enormous confounding background of potential aerosolized materials, also adds to the challenge. The high concentrations of agent that are likely required for standoff detection to provide a clear signal requires the dispersion of large quantities of material—something that is less likely to be the case in an asymmetric warfare situation.[11] In light of these challenges, it is clear that expertise must be obtained via collaborations between experts outside of the government with expertise and testing facilities within DoD and potentially other government agencies. Unexplored, but of potential promise, is adapting the multimodal architecture which evolved in countering improvised explosive devices to the chemical or biological detection problem (i.e., using more conventional ISR techniques that identify and track suspicious activities to queue forward deployment of chemical or biological detection systems).

---

[10] While it may be possible for DoD to acquire existing diagnostics from other government agencies, assay inadequacies due to age and issues related to the particular platforms used by other agencies make adoption difficult. No other agency is currently funding the range and hierarchy of modern diagnostics that the DoD requires for its worldwide missions.

[11] The committee is in general agreement that active point detection (e.g., sensors in a drone) may be more likely to achieve desired results within reasonable time/cost/error constraints.

Unknown Agent Identification and Characterization

Continuing human encroachment upon remaining wild lands worldwide, particularly in tropical or subtropical climatic regions, has contributed to a continuing stream of "novel" pathogen outbreaks in recent decades. Severe acute respiratory syndrome (SARS), Ebola, and Nipah are three biological examples of novel pathogens. Non-traditional agents are similarly unpredictable in the chemical domain. DoD requires that there be a robust (available, timely, and skilled) capability to identify and characterize unknown agents, both biological and chemical. This capability can be generalized as a hierarchy of detection approaches: [low resolution] presumptive field assays (rapid and inexpensive) that can rule dangerous agents in or out; [medium resolution] sensitive laboratory confirmatory tests for all known dangerous agents (biological or chemical); and [high resolution] robust suite of broad-spectrum analysis technologies to detect and characterize new, mutated, or engineered dangerous agents.

In this area today, DoD collaborates with many of the best researchers outside of government. However, it may well be necessary that DoD have an in-house capability so that classified samples can be processed. It is important that a robust identification and characterization capability for unknown agents be maintained. DoD has facilities and competent staff who should certainly be part of this capability. There are other non-DoD government agencies such as the Department of Homeland Security's (DHS) National Biodefense Analysis and Countermeasures Center (NBACC) facilities that have competence with unknown agent identification and characterization that can be considered part of the overall capability.

CBDP is currently sponsoring strong collaborations outside DoD for unknown biological agent identification and characterization. This collaboration involves scientists at DoD laboratories and is providing valuable training and technology transfer to their facilities. This successful model should be sustained and enlarged to include other key US participants as part of a robust US response to novel biothreats.

CHEMICAL AND BIOLOGICAL AGENT DETECTION

Detection of relevant chemical and biological agents is an important capability that underpins the CBDP. Sensors are needed that not only detect and identify agents over a wide range of concentrations, but also quantify airborne, waterborne, and surface contamination levels with adequate specificity to minimize false positives while responding as near to real time (seconds) as possible. The program has developed a wide range of point detectors with varying levels of sensitivity, specificity, and response times for a variety of relevant agents. In many cases improvements in sensitivity, specificity, response time, automation, robustness, size, weight, power consumption, and consumable reagent use would be beneficial. The development of standoff detection instruments with adequate sensitivity, specificity, response times, and multiagent capability has been difficult.

Systems that substitute real-time point agent detection deployed on unmanned air or ground vehicles may supplement or replace traditional remote sensing systems. Agent detection instrumentation systems that include automated sample collection, instrument control, data acquisition, and analysis and handling must be engineered for both laboratory and field use; and suites of instruments may need to be integrated with either static or mobile platforms for field deployment. Accurate agent transport analysis models to predict future agent distributions based on current agent distributions are also required.

The committee has identified four primary S&T capabilities that are required for the CBDP to achieve their Chemical and Biological Agent Detection objectives. The S&T capabilities are Analytical Methods Discovery, Instrumentation Development, Sensor Systems Development, and Agent Transport Analysis.

Analytical Methods Discovery

Current and anticipated chemical and biological agents must be characterized to identify and assess unique molecular and/or biological features that can be exploited to allow detection at acceptable levels of sensitivity and specificity in as near real time as possible. Chemical agent detection might be based on spectroscopic, mass spectrometric and/or chromatographic or mobility properties of the molecular agent or a derivative agent reaction product. For biological agents, detection may be based on DNA or other biomarker identification (e.g., protein, RNA). The capability to determine which structural and/or reactive agent properties can be exploited for reliable agent detection is the critical first step in developing reliable analytical agent detection methods. Subsequent experimental work can determine which identified analytical methods are reproducible, robust, and inexpensive enough that they could be developed into reliable laboratory and field agent detection technologies.

Analytical methods discovery is primarily a basic science endeavor in molecular recognition. This capability can be obtained in DoD laboratories, other government laboratories and in academic and private-sector laboratories. However, the ability to provide and safely handle the chemical and biological agents[12] necessary to test and assess analytical methods is reliably found in DoD facilities (including test ranges).

[12] The committee recognizes that there are many laboratories throughout the United States with the capability to work with up to Biosafety Level 4 (BSL-4) pathogens. These private laboratories, however, can be closed at any time for any reason (e.g., an accident at a privately owned BSL-4 laboratory may lead to immediate, and possibly permanent, closure).

DoD laboratories will often benefit by collaborating on innovative analytical methods development with leading academic or private-sector analytical methods development laboratories. A formal requirement for the laboratories to engage the extended analytical research and development community will likely increase the success of this S&T capability for DoD. Conversely, non-DoD laboratories will likely need to collaborate with DoD facilities for live-agent testing efforts.

Instrumentation Development

Promising analytical research methods capable of detecting and quantifying chemical agents need to be developed into reliable instruments capable of routine, and ideally automated, laboratory and field measurements. For detection of many biological agents, effective laboratory assays must be periodically re-evaluated to ensure that recently accessible strains are appropriately classified within the assay. The development of robust and reliable instruments requires a wide range of scientific and engineering skills that are often not all available in individual laboratories. Soon after proof-of-principle experiments have been successful, developers of analytical methods that may serve as the basis for innovative instrumentation should involve potential collaborators (especially potential end-users) to define mission-realistic measurement requirements and potential instrument specifications.

Since successful instrumentation development requires such a diverse range of scientific and engineering skills, collaboration is often required. While the basic analytical methodology may have been developed in the DoD, other government, academic, or private-sector laboratories, the full range of skills required to develop and demonstrate robust and reliable instruments are found predominantly either at selected government or private-sector laboratories. Further, cost-effective routine production of instruments is usually only available in the private sector.

Sensor Systems Development

Individual agent detection instruments have limited utility, especially outside a laboratory. To be fully useful, field instruments need to be integrated into a suite of instruments and deployed as a measurement system with automated sample acquisition, instrument control and data acquisition, quality assurance, analysis, transmission, and archiving capabilities. Laboratory instruments may not need to be as fully integrated. Agent-detection instrument suites may be deployed at fixed sites or on mobile platforms, including ground vehicles, ships, and piloted or unmanned aircraft.

Some sensor system development capabilities are available at DoD laboratories, other government and government-sponsored laboratories, and in the private sector and often require collaboration between these entities to acquire the required range of scientific and engineering skills. Systems integration with mobile platforms is often done by private-sector engineers. Instrument system testing and evaluation with actual agents usually require DoD facilities or collaboration.

Agent Transport Analysis

Understanding the present or future distribution of dispersed chemical or biological agent is an important capability for DoD personnel dealing with an attack. Airborne agents can be dispersed by the wind, as can evaporating or re-aerosolized agents from contaminated surfaces. In these cases atmospheric dispersion and meteorological models can be used to predict how fast and how far agent contamination might spread. Agent-contaminated surface waters can also be dispersed by stream flow and/or wave action and contaminated soil can be aerosolized as dust or be transported by vehicle tracks or tires. Precipitation can help transport agent into exposed soil. The development and testing of models to assess agent dispersal mechanisms and rates in various environmental media is a multidisciplinary challenge, and understanding these effects will have implications both for detection system and method development and the testing and evaluation of those systems. Biological agents can be dispersed by both the physical mechanisms listed above and by infected animal, bird, insect, and/or human vectors, creating additional challenges (see section on agent transport and viability analysis in Hazard Assessment, Management and Decontamination).

DoD laboratories have limited capabilities to develop or refine environmental agent dispersion or infection outbreak models. The Defense Threat Reduction Agency (DTRA) has recently funded significant research and development (R&D) activity to develop and test improved atmospheric dispersion models, including inverse models that predict agent release point and size from downwind concentration data. Most groundbreaking environmental dispersion research is currently performed in other US government laboratories (e.g., National Center for Atmospheric Research, Lawrence Livermore National Laboratory (LLNL), Sandia National Laboratory, Los Alamos National Laboratory (LANL)) or in academic or private-sector laboratories.

Many biological agents (excluding toxins) have a reproductive capability that differentiates them from chemical agents. This requires analysis of both agent fate, transport (effect on individual spores/cells/virions, etc. due to physical processes), and transmission via reproduction in hosts and/or vectors. Agent fate studies examine the impact of factors such as ultraviolet light, desiccation, and re-aerosolization upon biological agents at the individual level. Agent fate studies are performed in academia, in DoD and other government laboratories, and in industry. Agent transmission studies focus on population behavior of biological agents across large or small outbreaks that may span time scales far longer than those of individual agent particles. Such epidemiological modeling of infectious disease outbreaks is performed in academic centers, Department of Energy (DOE) laboratories, and in other government agencies (e.g., CDC and USDA). The CBDP should consider continuing to outsource most of their R&D in this area, specifically for biological agents, and focus on collaborative relevant live agent testing.

INDIVIDUAL AND COLLECTIVE PROTECTION

Deployed troops confronted or threatened with dispersed chemical or biological agents need both personal protective gear and collective protective shelters that will mitigate agent effects to the extent feasible. Personal protective gear may include respiratory masks and controlled-permeability suits, boots, and gloves. Collective protection may be afforded by air-locked and sealed temporary battlefield shelters with filtered air supplies, as well as by more permanent sealable, air-filtered spaces shipboard or at operational bases and other military facilities. Both personal and collective protective gear must be designed to minimize physiological stresses induced by their use, including heat stress, respiratory difficulties, excess weight burdens, cardiac stress, and sweat and waste accumulation.

The committee has identified five S&T Capabilities that support Individual and Collective Protection. These include Controlled Molecular Transport Materials Discovery, Barrier Materials Engineering, Personal Protective Systems Development, Collective Protection Systems Development, and Physiology.[13]

Controlled Molecular Transport Materials Discovery

Materials scientists can now design and characterize materials, fabrics, polymers, and construction materials tailored to restrict agent penetration while allowing passage of oxygen, water, carbon dioxide, and other substances that humans need to consume or emit. Robust, lightweight materials with controlled chemical and heat transfer properties

[13] Prophylactic medications and topical creams are covered in Medical Countermeasures below.

can be used to develop improved absorbents, fabrics, and other materials required for improved personal and collective protection gear. Materials can also be designed to trap or deactivate chemical or biological agents. The design of new tailored materials may greatly improve the effectiveness and greatly reduce the stressors associated with the use of current protective gear.

While DoD laboratories have some capability to achieve the fundamental materials science advances required to design and demonstrate new materials with tailored mass, heat transport, and agent deactivation properties, much of the nation's ground-breaking capabilities are found in academic and private-sector laboratories. The committee recognizes that significant collaborative efforts in this area are ongoing and continued collaboration with selected non-government laboratories is advised.

Barrier Materials Engineering

New prototype materials with controlled molecular transfer properties have to be engineered into fabrics, plastics, filters, and other robust materials that can be used to develop and test better protective gear. Properly engineered new materials with tailored mass and heat transport properties are the key to more effective and usable personal protective gear. This directly affects the mission capability of the warfighter.

DoD laboratories do have significant capabilities in barrier materials engineering, but collaboration with selected academic and private-sector scientists and engineers is advised to ensure in-house capabilities are regularly renewed and expanded. DoD personnel and facilities will usually be required to test new materials against relevant threat agents.

Personal Protective Systems Development

New material formulations have to be incorporated into prototypes of improved protective gear by teams of experienced materials scientists and engineers. DoD laboratories have suitable expertise and should continue to lead in the production of initial prototypes. Still, they should seek to collaborate with private-sector technical experts as they may be required to produce sufficient items for extensive test and evaluation activities. Skillful prototype development and extensive and *realistic* testing is critical to achieving improved personal and collective chemical and biological (CB) protection products.

Given the current lack of commercial markets and the relatively small size of the DoD market for CB protective gear, limited prototype engineering capability is available outside DoD. In addition, DoD personnel and facilities will usually be required for extensive prototype testing.

Collective Protection Systems Development

Collective protection systems development involves some additional considerations compared to personnel individual protection. New material formulations have to be incorporated into prototypes of improved collective protection systems (e.g., shelters) by teams of experienced materials scientists and engineers. Shelter materials must have structural strength as well as agent protection properties. Skillful prototype development and extensive and realistic testing is critical to achieving improved collective CB protection systems. These systems directly impact warfighter survivability in both base and forward environments.

For the reasons discussed in the personnel protection systems development section above, limited prototype engineering capability is available outside DoD and close collaboration with the few commercial suppliers is needed.

Physiology

The ability of individual protective gear (e.g., suits, masks, gloves) to permit operations in the presence of chemical or biological contamination must be weighed against the effects of the protective gear (e.g., increased heat load, increased moisture retention, decreased manual dexterity and overall maneuverability, potential psychological effects, etc.). Expertise in human physiology is required in all stages of design, development, testing, and operational transitioning. To a somewhat lesser degree this also can apply to collective protection (e.g., shelters).

It is imperative that the warfighter be provided with effective individual and collective protective solutions, and access to skilled human physiology expertise is extremely important in the design, testing, and deployment phases. Basic human physiology expertise can be found outside of the DoD and government. To be maximally relevant to CBDP, extensive on-site collaboration with the designers, testers, and ultimate end users of the protective gear is mandatory for any physiology collaborators.

MEDICAL COUNTERMEASURES

Development of countermeasures, both prophylactic and therapeutic, is vital to provide the warfighter protection against either biological or chemical agents. The S&T capabilities required for development of these countermeasures have increased in complexity[14] due to the requirement

[14] The complexity arises from the need to perform efficacy testing of a candidate compound. Because of the inability to perform this testing in human subjects, the FDA Animal Rule for efficacy has been introduced (vida infra).

that the DoD acquire FDA licensure for products given to military service members. Many of these capabilities should be coordinated with ongoing activities with NIH and the Biomedical Advanced Research and Development Authority (BARDA), agencies that are developing CB countermeasures for the civilian populations and have many overlapping capabilities required for countermeasure development that can be leveraged. The committee found eight S&T Capabilities that are needed to achieve CBDP objectives in Medical Countermeasures: Target Discovery, Regulatory Science, Mechanisms of Delivery and Delivery Systems, Animal Models, Host Response, Pre-Clinical Studies, Clinical Trials (GLP and GMP), and Medical Product Development.

Target Discovery

Identification of critical targets for the development of new vaccines, antibiotics, diagnostics, and other pharmaceuticals are required for the development of countermeasures to protect the warfighter. Continued target identification for new and existing DoD-relevant biothreats will be critical for development of new innovative countermeasure drugs, vaccines, and potentially new therapeutic platforms.

In general, and particularly for biological agents, there are many outside sources that can provide this service either within the pharmaceutical or biotechnology industries or academia. The caveat is that this capability may be limited by the biosafety level required for the development of the targets. For example, if target development or antibiotic screening requires live, virulent agents and it is a BSL-4 agent this significantly limits where the studies can be performed. In many cases initial screens have been performed and reported that can be leveraged by the DoD laboratories for licensing and further development. The most productive route is likely through collaborations with industrial, and in some cases academic, partners willing to apply their target identification strategies to pathogens with high DoD relevance.

Regulatory Science

Over the last decade there has been recognition by many of the FDA regulatory challenges involved in countermeasure product development due to the multiple deficits in our basic understanding of many of the biothreat diseases and the application of the "Animal Rule"[15] to the process. The acknowledgment of the impact of the FDA requires a product development process that embeds an element of regulatory science with

[15] 12 CFR Parts 314 and 601.

appropriate expertise related to the application of Good Laboratory Practice (GLP)-based assays and protocols. In many cases the implementation of this capability requires a significant addition to current infrastructure and a shift in the research culture at appropriate institutions. Full understanding of key elements of regulatory science will be critical for the successful development of countermeasures. The committee was informed that DTRA is now supporting an FDA effort to establish a new genome reference database to accelerate the validation process for highly multiplexed assays utilizing modern technologies. This is a laudable example of interagency collaboration.

There are multiple contract research organizations that are well prepared and have the capabilities of applying GLP work for product development up to BSL-3. While this covers the vast majority of potential biothreats, BSL-4 biothreats and some chemical agents, a number of other biothreats, and chemical agents may require DoD expertise and facilities. Maintenance of these latter capabilities focusing on specific high-priority pathogens and chemical agents should be maintained for test and evaluation (T&E).

Mechanisms of Delivery and Delivery Systems

This set of capabilities is concerned with being able to deliver pathogens, drugs, and vaccines by multiple routes that will mimic likely exposure and therapeutic delivery modalities. For example, one clear modality of concern to DoD is aerosol delivery which can be used in exposure protocols or as a strategy for therapy delivery to a target organ.

It is vital for model development to maintain expertise in delivery strategies that are likely to be the route of exposure for human populations. Aerosol delivery has been the focus for decades, but all routes need to be considered (e.g., food may be seen as a bioagent delivery mechanism by terrorists). Further, progress in therapeutics is occurring not only by breakthroughs in target identification and chemical modifications, but also in delivery mechanisms. For example, the use of nanoparticle-based delivery mechanisms is being investigated by both pharmaceutical companies and basic research scientists worldwide. The capability to perform the delivery may be as important to CBDP as studying the basic science behind the delivery. Strong connections to appropriate institutions that perform this work should be developed so that DoD personnel can be trained and current infrastructure updated as needed.

Animal Models

Direct human testing will not be possible with most, if not all, biothreat agents to allow for the evaluation of countermeasure efficacy in humans. Therefore, the capability to develop high-quality animal models of infectious diseases in multiple species is critical for countermeasure development. Animal models against chemical agents and toxins are required for similar reasons. The maintenance of this capability requires multiple disciplines including pathology, immunology, systems biology, and aerobiology. The latter skill is required in order to accurately mimic the most likely exposure routes that military personnel will face. These models should be readily available to rapidly assess new antibiotics or vaccines that are developed, whether developed by DoD or elsewhere.

The "Animal Rule" requires that sponsors of products in development demonstrate efficacy in at least two animal species that accurately represent the infectious disease in humans. In some cases, this requirement is extremely difficult because of the limited knowledge base related to some of the infections (e.g., viral hemorrhagic diseases) most relevant to military personnel. Thus, it is important to remember that there remains a significant amount of basic pathophysiology that needs to be performed in animal model research as well as efficacy testing.

This capability is one that can be achieved through multiple routes as there are multiple outside parties that can perform both proof-of-principle and GLP models in multiple species, including primates. Furthermore, HHS, including NIH and BARDA, has invested in academic institutions to develop capabilities in model development for a variety of biothreats. Many universities have now developed the infrastructure for producing reports compatible with product development. While there is now broad development and use of animal models in academia and industry, the DoD laboratories needs to maintain, at least in part, this capability for pathogens and chemical threats that require highly specialized (bio)containment facilities to ensure availability.

Host Response

Ordinarily this would be a component built in to the animal model development phase; however, there are specific elements of the "Animal Rule" that make this capability of critical importance particularly as it relates to vaccine development. Thus, the "Animal Rule" requires that the protection elicited by a countermeasure in an animal model must mimic a similar type of response in the human. Typically this is referred to as "correlates of protection." While this is more easily proven for antibiotics and chemical agents, vaccines generate host-driven immune responses that can be quite variable and unpredictable among species. Therefore, understanding the mechanisms of protection in the animal models, as well as developing innovative strategies for collecting as much information regarding the mechanisms of vaccine protection by humans, is critically important for countermeasure development.

Since the correlates of protection are a major milestone for countermeasure development through the "Animal Rule," this capability is critical for countermeasure development. It is important to realize that the anticipated design of procedures needed to meet the Animal Rule requirements shall begin early in the R&D process as it may change how pivotal experiments are performed.

Since at this point there is no roadmap for developing correlates of protection, this process is likely to be a multidisciplinary approach requiring multiple collaborations with agencies that can provide strong computing power and relevant data from *in vivo* and *in vitro* assays. Such strategies are probably best performed via strong project management coordination with funded academic centers and other agencies.

Pre-Clinical Studies

This capability encompasses a broad range of activities including GLP studies. Such activities include proof-of-principle studies for efficacy, dose-ranging studies, pharmacodynamics and pharmacokinetics, and toxicology. These activities are critical to down selection of new therapeutics. Optimally, these would be performed in a head-to-head fashion to compare potential products for further development.

Some of these studies are small in size and scope, therefore, lending themselves to outsourcing. Except in cases where the BSL category would limit its location, these studies could be performed inside or outside of DoD. The construction of the new BSL-4 at the University of Texas Medical Branch provides an example of a conduit for performing even high biocontainment agent studies outside of DoD. The new DoD laboratories will have expanded capacity to do this work as well.

Clinical Trials (GLP and GMP)

While efficacy trials for threat agents will not be performed in humans, the FDA still requires appropriate human safety trials. This capability will be required for advancement of drugs beyond animals, and requires access to appropriate infrastructure to handle human clinical trials (i.e., institutional review boards, informed consent, and safety monitoring). Included in this activity is post-marketing surveillance (phase 4) where if the product needs to be utilized, there is an infrastructure in place to monitor outcomes regarding safety and efficacy and potential delayed effects during large-scale use.

Advanced development and acquisition of FDA licensure will require multiple stages of clinical trials. Demonstration of safety is a critical component, not only to the FDA, but also to the military personnel who will be receiving the new pharmaceutical.

This capability has been performed for decades outside of DoD by medical schools and agencies that perform this activity under contract. The DoD should leverage the external capabilities (within or outside of government) available for clinical trials rather than replicate it.

Medical Product Development

Methodologies of product production have evolved over the last decade. Because of the evolving nature of the threat, the ability to rapidly scale up manufacturing to provide safe and effective countermeasures for military personnel is needed. This capability would allow the military to have flexibility with regard to its manufacturing strategies and allow for rapid implementation of new product production without major delay.

This capability is likely best found, or developed, outside of DoD with collaborations with academia, industry, or combinations thereof where the experience of product development and appropriate personnel are maintained.

HAZARD ASSESSMENT, MANAGEMENT, AND DECONTAMINATION

In the DoD context, hazard management is basically limited to avoidance, quarantine, or decontamination. This section will focus primarily on the S&T aspects of decontamination. Hazard assessment has two aspects: detection of the hazard (which was covered above) and determination of how well the hazard has been decontaminated, which is discussed here. In order for the CBDP to address the area of Hazard Assessment, Management, and Decontamination the committee identified five S&T Capabilities that are needed by the CBD Program. These include Decontamination Methods Discovery, Decontaminant Development, Decontamination Resilient Materials Development, Decontamination Systems Engineering, and Agent Transport and Viability Analysis.

Decontamination Methods Discovery

Methods discovery for decontamination concerns the research and development to identify what will neutralize or kill chemical or biological agents. Ideally, decontamination methods will be effective against a broad array of potential agents and pathogens. Also, it is desirable that methods will not be corrosive or otherwise destroy the functionality of materials and equipment being decontaminated.

Without effective decontamination methods, the only choice will be to sequester any material, equipment, or area that has been contaminated. For some agents (e.g., anthrax spores which can survive for decades) this could effectively mean near-permanent denial of use.

Methods discovery can and does occur outside of the government. CBDP should partner with the best expertise available to obtain the needed methods discovery. However, testing of the effectiveness of any method can and should be done with real agents in a DoD facility. Testing at scale for any decontamination methods requires the use of extremely focused DoD facilities.

Decontaminant Development

In order to be an effective decontaminant, a substance must be capable of being applied in an appropriate manner, be effective in real-world environments where surfaces will not be in a pristine condition, and result in an ability to be functionally restored (e.g., electronics are not destroyed).

Without decontamination substances there will be a lack of ability to continue missions in both the near and long term. While fighting contaminated might be effective for an individual campaign, it is not effective for a war. Having effective decontamination substance(s) available when necessary is essential for the warfighting equipment and for all supporting equipment. It is also necessary to have wide-area decontamination substances and delivery methods available for battle spaces, and especially necessary for civilian areas. Appropriate consideration of concept of operations (CONOPs) will enable DoD to develop different decontaminants to address different operational needs (i.e., decontamination of a military aircraft in a campaign is different than decontamination of a building).

To obtain this capability, CBDP should partner with the best expertise available to enable the needed decontaminant development. However, testing of the effectiveness of any substance can and should be done with real agents in a DoD laboratory or facility.

Decontamination Resilient Materials Development

The development of materials that can withstand decontamination processes, or even self-decontaminate (e.g., paint or fabric embedded with neutralizing additives), is needed for warfighting and support equipment. Such materials would survive, with full functionality, a standard low-cost decontamination process or would not need to be decontaminated at all. Resilience is especially important for high-value, sophisticated equipment such as electronics in aircrafts, or for porous materials, such as painted surfaces or seat cushions, that are quite difficult to effectively decontaminate.

In the absence of developing materials that are resilient or self-healing, decontamination substances and systems will require much more sophistication and time to develop. Resilient material development could greatly reduce the time, cost, viability, and effectiveness of decontamination.

Resilient materials development is an important capability that can be advanced in parallel with other decontamination strategies. The CBDP should leverage outside sources that are already working on this problem. Rigorous testing of the effectiveness of any material can and should be done with real agents in a DoD governmental laboratory or facility.

Decontamination Systems Engineering

The decontamination system includes the means of delivery of a decontamination substance, the logistics for the decontamination, i.e., moving the material to the proper location, cleanup after the decontamination process, and an ability to determine the decontamination has been effective.

There are two types of systems: (1) material and equipment and (2) wide-area battlefield, containment area, or urban area. For material and equipment, the logistics are quite different and easier to develop and implement than those for a wide area. In a wide area there likely needs to be decision aids to help determine the priorities and methods (a completely different type of CONOPs from material or equipment) for effective decontamination. In all cases, the tracking of contaminated substances to other areas and the potential for re-suspension of materials must be considered.

Decontamination substances alone will not be enough to fully utilize and maintain warfighting capabilities during wartime. Without the systems decontamination ability, decontamination will not work and material, equipment, or wide areas will simply need to be declared unusable.

Decontamination systems engineering can and does occur outside of the government. CBDP should partner with the best expertise available to obtain the needed decontamination systems engineering. However, testing of the effectiveness of any engineered system and getting feedback from operators on the process and decision aids will be essential. While the ability to test decontamination systems on real equipment exists within the DoD laboratories and facilities, the reluctance (due to overall cost of test and decontamination) to actually conduct the tests seemed very high, thereby making them unavailable for full-scale, live-agent testing.

Agent Transport and Viability Analysis

Decontamination of biological agents is highly agent specific. Some biological agents are quite fragile and will only survive a brief exposure to the environment (e.g., ultraviolet light kills many bacteria) while others have extremely robust survival mechanisms (e.g., anthrax spores can survive for decades). Many robust operational molecular detection methods for biological agents are based on nucleic acids and will correctly detect both live and intact DNA as well as dead non-intact DNA—without distinguishing between the two. Thus, robust, timely *viability assays* are important for successful hazard assessment, management, and decontamination. CBDP needs to be able to deliver robust viability assays for a wide range of potential biological contaminants. Methods for characterizing chemical agent contamination on either military equipment or environmental surfaces also require continued development. Recent advances in ambient ionization mass spectrometry methods provide real-time surface contamination measurements that can determine distributions of agent contamination both before and after decontamination treatments.[16]

To the DoD, a critical question after a biological contamination incident will be "Is it safe now for people to remove their protective gear and resume normal operations?" Thus, the ability to provide robust viability assays for any biological agents that could be reasonably anticipated is of high importance to CBDP. Of equally high importance is the ability to rapidly create robust viability assays for any additional biological agents that were not anticipated or known in advance of their use.

As with detection assay development, viability assay development can and does occur outside of the government. CBDP should partner with the best expertise available to obtain the needed viability assay capabilities for program specific needs (i.e., threat agents). Ambient ionization mass spectrometry systems that can be adapted to characterize chemical agent contamination of surfaces is now available from several commercial sources.

CROSS-CUTTING SCIENCE AND TECHNOLOGY

Numerous critical aspects of CBDP underlie many of the specific capabilities discussed above. These foundational capabilities are discussed in this last section; however, they should be recognized as being of crucial importance to each of the operational capability sections discussed above. It should also be noted that many of these infrastructural capabilities are

---

[16] The National Research Council Committee on Assessment of Agent Monitoring Strategies for the Blue Grass and Pueblo Chemical Agent Destruction Pilot Plants recently completed a study in this area. Their report can be found online at www.nap.edu.

assumed to be fully available by other program elements for their use (often at no cost or at less than full-cost recovery), yet funding for infrastructure may have dropped below critical mass in some cases. The thirteen cross-cutting issues identified by the committee include Acquisition, Maintenance and Transport of Critical Chemical and Biological Reagents; Simulation; Informatics; Forensics; Education and Training; Behavioral Analysis; Systems Analysis and Engineering; Repurposing Commercial Technologies; Systems Biology; Synthetic Chemistry and Biology; Materials Science; Statistical Measurement Design; and Test and Evaluation.

Acquisition, Maintenance, and Transport of Critical Chemical and Biological Reagents

CBDP needs to be able to acquire or generate, maintain, and transport all chemical and biological reagents necessary to support the development and testing of detection, characterization, and viability assays. This includes traditional chemical agents, non-traditional agents, and biological agents and toxins, as well as all other relevant agents needed for test panels and other research, development, test, and evaluation purposes. Lack of required reagents in a timely fashion can slow or seriously derail nearly all major CBDP research, development, test, and evaluation programs. This cross-cutting S&T capability supports nearly all other programs within CBDP.

Certain aspects of both chemical and biological reagent creation, use, and transportation are severely limited by US and international law and conventions.[17] There are very few facilities permitted to create or work with threat agents. Similarly, there are relatively few facilities where BSL-4 Select Agent use is permitted. In the case of Select Agents, there are facilities outside of the government that are permitted to create and work with these agents. There are numerous facilities that are permitted to work with BSL-3 biological agents (including relevant Select Agents).

The current DoD Critical Reagents Program maintains a robust capability to grow, maintain, and transport biological reagents. This includes live agent, DNA/RNA, and antibodies. Similarly, the ECBC is where controlled chemical agents are manufactured and transported for DoD research and live-agent testing needs. The DoD is a primary source of capability for live-agent production and testing of both chemical and biological agents, particularly in terms of the scale of the testing facilities.

---

[17] The Chemical Weapons Convention (CWC) permits member-states to operate a single small-scale facility for the production of chemical agents that are included in the schedules of the CWC. The Edgewood Chemical Biological Center (ECBC) is currently the designated facility in the United States.

Agent Simulation

Many of the chemical agents cannot be tested in full-scale outside exercises. Information about them must be inferred from "simulants"—that is, compounds that closely approximate the properties of the molecule in as many properties as possible *other than* toxicity—or by "simulation"—that is, computing the behavior and characteristics of the compound from study of the behavior and characteristics of compounds of similar structure. Although computer-based simulation has developed extraordinarily rapidly, accurate simulation still depends on a foundation of empirical knowledge obtained through experiment. The DoD laboratories are uniquely equipped to carry out calibrating studies with actual agents in secure laboratories; but some types of tests require large-scale experiments, or experiments under a variety of conditions that cannot be simulated accurately in the laboratory.

The state of development of chemical simulants could be substantially improved. A number of compounds have been examined, but work in this area has been limited for reasons unrelated to the science: (1) simulants do *not* replicate the properties of the agents; (2) even if good simulants were developed, their use would require complex environmental protocols to demonstrate acceptable environmental impact; and (3) they might be expensive. These statements may well be correct, but given the importance of the problem, they are *not* acceptable reasons to neglect the development of accurate simulants. This activity is one that has the potential to be the basis for a good collaboration with university laboratories, with the universities combining physical organic chemistry, synthesis, and computer-based simulation of properties, and the DoD laboratories carrying out comparisons with active agent. Alternatives have to be developed based on considerations of structures. It is important to emphasize that a simulant does *not* have to behave indistinguishably in all respects from the chemical agent to be useful. Thus, a compound that would simulate permeability through fabric would not have to replicate volatility; what is required is that the differences are known and can be corrected for.

With these points of calibration, it should be possible to build more elaborate and predictive computer models for dispersal, migration, environmental deactivation, skin permeability, and so on; the technology for such models has developed very rapidly in the last years, but has not permeated the DoD laboratories concerned with CBD, other than in some areas of dispersal and related hydrodynamic issues.

It is critical, in the committee's view, to pursue a program in simulants and simulations aggressively; not to do so risks making claims for effectiveness of technology and doctrine that may not hold up in conditions that have not been tested.

Informatics

Biology has recently matured to become an information-dominated science. Chemistry has been in that state already for many years. Both bioinformatics and cheminformatics are essential underpinnings for all research and most development efforts in CBDP. Bioinformatics is a broad term that can be used to cover both infrastructure (e.g., sample tracking, laboratory information systems, web sites, and databases) and more domain-specific computation (e.g., genomic comparisons, protein structure modeling, and pathway analysis). Similarly, cheminformatics can describe, for example, atomistic modeling, small molecule docking, and drug compound *in silico* screening.

The existence or non-existence of sufficient informatics capability can literally make or break critical projects and programs in chemical and biological defense. In the biological domain, information is undergoing exponential growth driven by a recent 1,000 times increase in DNA sequence data production rates, and informatics will become an increasingly important tool for CBDP in the coming years.

Currently most of the informatics capabilities are found outside of government proper. Some DoD laboratories have made large strides increasing their in-house capabilities in the past few years, particularly in terms of *de novo* annotation of bacterial genomes. It appears that increased collaborations with outside experts would be the most efficient way to ensure that CBDP programs are receiving the required informatics expertise in both the chemical and biological domains.

Statistical Measurement Design

Proper experimental design underlies testing and evaluation of all CBDP products and systems. How many times and at what concentrations of agent should a new piece of protective gear be tested? What percentage of production lots should be tested to ensure high confidence that those lots can be delivered to warfighters for effective use? Rigorous statistical expertise should be applied to both design the right number and nature of experiments as well as to accurately assess the results.

The lives of warfighters are at stake for virtually every product that results from a CBDP. The proper design and analysis of test experiments is therefore of very high importance to CBDP.

Statistical testing expertise is available at multiple locations outside of government. The committee observed evidence that there could be greater use of external expertise to review testing plans that are currently created internally and apparently not subject to independent external review (see also Test and Evaluation).

Forensics

An important goal of CBDP is detecting and mitigating chemical and biological threats to the warfighter. A secondary goal, however, is to develop answers to key intelligence questions related to the perpetrators, origin of the source materials, and how the biological or chemical weapon was produced and disseminated. For example: How did a food poisoning event transpire? Was anthrax used on US or allied forces from a natural exposure in contaminated soil or was it engineered in some fashion to be more lethal? Was it a strain one would expect to find in that location of the world? Could it be associated with a strain that has been or could be identified as having originated from a known facility? Could any chemical signatures in the agent used on US or allied forces indicate what production process was used, who might have the capacity and competency to produce or disseminate agent, and who might have aided the perpetrators?

Chemical and biological weapons forensics may leverage many of the techniques and technologies used for detection diagnostics, plus other techniques that examine orthogonal dimensions (e.g., isotopes in the water used in manufacturing might provide a clue to origin). CBDP has invested in some of these activities. For example, DNA sequence analysis may drive both the development of detection diagnostics and supply evidence of potential genetic engineering, which is of forensic value. For several reasons, the DoD chemical and biological laboratories may be involved in attempts to augment other US government capabilities to perform forensic analysis of incidents affecting US forces in order to inform attribution decisions and mission planning and decision making.

The committee is aware that chemical and biological forensics is not currently in the purview of CBDP, but is handled by other DoD and DOE elements. Chemical forensics is handled primarily at two sites (ECBC and LLNL). Bioforensics is presently in a more nascent form. Greater synergy between the CBDP, DHS, and intelligence community programs, including the Defense Intelligence Agency, could prove useful to improve overall DoD diagnostics and forensics capabilities.

Education and Training

Implementation of capability in CBD will usually have components in both technology and training in use of the technology: the latter may be the more difficult, but depends on the former. Ease of use is a key concern in the development of most CBD programs.

The development of aids for education and training has been a long-standing interest in the DoD, with programs such as SIMNET and the Medical Management of Biological Casualties "Blue Book" being pioneering efforts. Adapting computer-aided programs from a wide variety of apparently unrelated activities (e.g., for sonar operators, helicopter pilots, and sniper detection) could provide useful methodology at relatively low cost. The development of realistic training protocols, with long-lasting impact, is a more complex subject. Since technology for CBD cannot be tested and learned in a "real" environment (as in live-fire exercises), and since a serious CB attack, or even a threat of one, produces great confusion and misuse of technology (as judged from experiences in the early days of the Iraq wars), the development of adaptable, durable, robust training protocols for users is an important issue, but one from which there is little established technology.

Behavioral Analysis

Earlier we discussed how physiology was important to study the impact of chemical and biological protection on the physical ability of the warfighter to function in protective gear. Similarly, behavioral analysis is important to understand the mental state of those required to perform their missions under the added mental and physical stress of an impending or actual chemical or biological attack.

Even if individual and collective protection gear functions perfectly as designed, there may be individuals who increase their risk of exposure due to behavioral factors (e.g., claustrophobia, extreme irritability due to discomfort, etc.). It is important for CBDP to learn as early as possible in the design, development, and test cycle whether new equipment has any characteristics that may increase the likelihood of such behaviors. A thorough understanding of behavioral factors may also provide useful design requirements.

Most human behavioral expertise resides in academia. Close collaboration between academia and DoD researchers, product developers, and operators will be needed to effectively translate this expertise into a useful outcome for the DoD.

Systems Analysis and Engineering

System analysis refers to an overall analysis of the various alternative means of meeting the mission requirements. Cost/benefit, threats, CONOPs, manufacturability, human factors, and behavioral analysis are all capabilities needed to conduct an appropriate systems analysis. Systems engineering refers to the detailed examination of how the totality of an individual system is likely to perform in its operational environment. Computer simulations are a critical part of systems engineering and include many factors of informatics.

A system analysis performed early in the R&D process can often eliminate technically interesting but operationally inadequate solutions. Integrating end-user expertise into a system analysis can lead to discovery of entirely new solutions to mission requirements.

Without systems engineering, the performed solution at best may be overly expensive, and at worst will fail to meet operational requirements.

System analysis and engineering are critical to CBDP, especially when incorporated into the process at early stages and updated throughout the RDT&E process. High-quality system engineering and analysis capability exists both inside and outside the government, although it does not seem to be resident throughout the CBDP.

Repurposing Commercial Technologies

The committee encourages the entire CBD community to take an active approach to following scientifically related fields of R&D and product development in an effort to identify non-CBDP projects, products, and personnel that may aid the CBDP meet its mission without direct, or with reduced, investment and shorten time to solution. Most of the activities involved (literature and public press tracking, active engagements in societal meetings, and engagement of fellow US government colleagues) are likely performed already by nearly every CBDP scientist and manager. Additionally, actively engaging pre-competitive alliances (particularly in the pharmaceutical industry) could provide the CBDP with sufficient direct awareness of the research-of-interest and partnership connectivity to justify the relatively modest membership fees.

In the procurement-type strategy much of the innovation for the required DoD capability or product is developed during the early stages of research and then subsequently transitioned into a scale that is needed. In some cases however, due to the rapidity of technology development, such as the materials sector and medical research, there can be a significant breakthrough that could rapidly be incorporated to provide a needed product.[18] The procurement mechanism lends itself toward the development of "blinders" that may mask the incorporation of concurrent innovative solutions to ongoing needs by limiting the ability of incorporating new concepts or emerging technologies into the process. The incorporation of a "tech watch" concept into the existing practice would have two elements, (1) mechanisms for searching and identifying relevant breakthroughs in the literature and private sector and (2) mechanisms

[18] As an example, the National Cancer Institute (NCI) has recently issued a solicitation for testing their older and unused drug candidates for new purposes. In this case, AZT was in the NCI drug archive and later found new use as the best therapeutic for HIV/AIDS.

and processes in place for incorporating the innovation into a T&E for the capability needed.

The DoD researchers might help focus the off-label requirement and specify the need for detectors with new assay capabilities (and contribute to the new assay development itself). It is not clear that a major role exists for the DoD laboratories to run large-scale repurposing panels or to perform detector repurposing development. The culture of program managers and scientists within the DoD should dictate that they are "smart buyers" first before a major RDT&E investment is made.

Systems Biology

Systems biology studies how the many different networks or systems within one or more living organisms interact. One example might be how proteins from some viral pathogens spoof their way past human immune defenses and hijack human proteins in order to accomplish replication of the attacking virus, with a byproduct being human illness. By definition, systems biology requires a complex collaborative interconnection and integration of diverse sets of knowledge and expertise. One hope for systems biology is the future ability to automatically analyze pathways in multiple pathogens (and their corresponding responses in hosts) to determine potential broad-spectrum medical countermeasures. (Note that this was a goal of the former Transformational Medical Technologies program.)

Systems biology is still very early in development, and most systems biology efforts in the pharmaceutical companies are conventional drug development programs that have been extended to include understanding of pathways rather than knowledge of single protein targets. Systems biology *requires* a strong base in fundamental science to be useful; it is *not* a silver bullet. Since chemical agents—especially nerve agents—attack multiple pathways, the topic is an ideal candidate for study in a systems biology program.

Systems biology is being performed worldwide throughout the various biology communities. CBDP should be able to leverage much of this work. However, it appears probable that not all of the desired aspects of systems biology related to pathogenesis will be supplied by the academic research community or other government agencies. Thus, it is likely that CBDP will need to be selective in supporting necessary systems biology efforts related to key mission needs.

Synthetic Chemistry and Biology

Synthetic chemistry is now capable of synthesizing almost any small- to medium-sized molecule. Chemical synthesis is no longer rate limiting for chemical and biological defense.

"Synthetic biology" is a name given to the rational, biology-based synthesis of compounds (small molecules or large) that requires manipulation of synthetic pathways using metabolic and genomic tools. The ability to use synthetic biology to recreate viruses and to create novel bacterial platforms should indicate to DoD that a useful fundamental capability is to be able to detect and characterize the application of synthetic biology applied for both beneficial and nefarious purposes. The timeline of when full mastery of synthetic biology will be achieved is unclear, but recent rapid advances in other aspects of biotechnology make it appear *imprudent* to suggest that massively engineered organisms will *not* be a potential threat in the near future, whether due to design or unintended consequences of a beneficial intention.

Materials Science

Materials science has a role to play in many of CBDP's R&D endeavors. Areas as diverse as temporary building construction materials to nanomaterials for targeted drug delivery and agent-surface transport modeling to studies on the degradation effects of decontamination methods on textiles draw on the field. As a result, it is important that the DoD maintain at least a limited in-house infrastructure to perform research in materials science and to be available to consult and collaborate with scientists as needed across the program. This is also an area, however, where its very ubiquity has resulted in a great deal of expertise within the government and commercial sectors. Many examples can be found in the field of nanotechnology, which has received heavy investment over the past decade. To leverage these resources effectively, it will be necessary for CBDP to encourage collaboration and engagement with those broad communities.

Test and Evaluation

The materials and space required, access to live agents, and knowledge of the operational realities and environments in which the warfighters function make test and evaluation a core S&T capability for CBDP. However, based on limited exposure to the program, the committee infers that T&E is an area that requires serious strengthening in areas relevant to CBD. Upon examination of the T&E structures in place, a number of questions emerge as difficult to answer: (1) Under what conditions of actual use (a combat soldier, in MOP gear, with weapons, pack, ammo, comms, under fire, in an environment with mud, thorns, rocks, sweat, and water) are the suits protective, and against what agents? (2) How, exactly, are these evaluations carried out (and they *must* incorporate simulants or "hazards" that are already environmentally common, e.g., diesel smoke as a particulate, poison oak/ivy as ground cover, fluorescent particles in dirt in an alley, and food dye in the wet sand of a landing zone). Test and evaluation protocols should be designed with consideration of the CONOPs the gear is intended to support. Data collected under controlled test conditions (and even the test conditions themselves) must be evaluated to aid in estimates of real world mission consequences and effectiveness.

The point of urging much more realistic and demanding T&E is to encourage development of a culture in which T&E produces accurate evaluations of how protective equipment, detectors, and operational doctrine function under conditions of use in conflict. This applies not just to the development of new equipment or materials, but also to currently fielded equipment that may have to protect against new threats. For example, established CONOPs may need to be revised if fielded equipment does not provide the same level of protection against an emerging agent as against the traditional agents. In order for a commander to plan and execute operations, it is important that test results can be collected, evaluated, and presented in a way that provides accurate information regarding possible casualties under a variety of conditions (lightly or heavily contaminated, damp or dry environment, etc.) and any restrictions or limitations on mission-critical activities that protective gear may introduce.

SUMMARY OF CBDP CORE CAPABILITIES

The committee identified 39 core chemical and biological defense S&T capabilities and created a framework that groups them in six categories. Using the decision framework discussed above the committee found that almost all of the capabilities **can** be found outside of the service laboratories. For each capability, R&D and T&E are discussed separately and typically were not best suited to the same organization.

The committee considered four types of institutions with laboratories that may be suited to provide CBDP core capabilities and organized them from typically having the most fundamental-science-focused to the most product-focused research. These institutions are (1) academia, (2) other governmental facilities (e.g., NIH, CDC, DOE National Labs, NIST), (3) DoD laboratories and facilities, and (4) industry (e.g., pharmaceutical companies). For some capabilities, T&E requires use of actual agent;[19]

[19] Actual agent testing refers to the actual chemical or biological agent the capability is being tested against (e.g., Vx, Sarin, sulfur mustard, anthrax, tularemia, botulinum toxin), as opposed to testing with simulants.

institutions other than DoD laboratories may be well suited to do the work but would need to do so in close collaboration with DoD.

Table 3.2, which summarizes the committee's judgments about how well suited the types of institutions are for R&D and for T&E with respect to 26 of the core capabilities, is reproduced below. Dark shades indicate an institutional category that the committee views as well suited to maintain a given capability for the CBDP, while the lighter shade indicates less well-suited locales. The white boxes indicate that the institutional category is, in the committee's view, not well suited to maintain the capability. The other 13 capabilities are cross-cutting science and technology that the committee views as necessary for effective RDT&E for any of the capabilities defined in the preceding capability categories. Discussion of the potential locales for the cross-cutting science and technology capabilities were discussed previously in this chapter.

The committee does not intend to imply that each of the thirteen cross-cutting capabilities be maintained exclusively, or indeed at all, within DoD.

When considering the various locales for obtaining S&T capabilities, it is important to recognize that

1. the analysis of the various laboratory locales is general, and individual performers within a category may be exceptions;
2. the color coding of each category represents the aggregate of reasons considered, including but not limited to
   a. reputation and experience at providing the given capability,
   b. the extent to which the capability requires work with classified information,
   c. limitations on the locale of the capability resulting from international treaties or other laws, and
   d. the need to maintain important capabilities, at least in part, at government facilities to ensure availability (e.g., BSL-4 facilities).

FINDINGS AND RECOMMENDATIONS

In identifying the science and technology capabilities necessary to support the Chemical and Biological Defense Program, the committee identified the following principle findings and recommendations.

Scientific Collaboration

Finding 3.1: Little of the *fundamental* science required for CBD lies *primarily* in the DoD. The vast majority of the scientific research performed in the United States occurs in academic and industrial laboratories. This is

Table 3.2: Science & Technology Core Capabilities

	R&D Academia	R&D OGF	R&D DoD	R&D Industry	T&E Academia	T&E OGF	T&E DoD	T&E Industry
1. Enabling CBRN ISR								
Information Acquisition & Analysis	well suited	well suited	less well suited	less well suited	not well suited	well suited	well suited	less well suited
Health Monitoring	well suited	very well suited	well suited	not well suited	less well suited	well suited	well suited	less well suited
Environmental Monitoring	well suited	very well suited	less well suited	very well suited	not well suited	less well suited	very well suited	less well suited
Unknown Agent Identification and Characterization	very well suited	well suited	well suited	less well suited	well suited	well suited	well suited	less well suited
2. Chemical & Biological Agent Detection								
Analytical Methods Discovery	very well suited	less well suited	less well suited	well suited	not well suited	not well suited	A	less well suited
Instrumentation Development	less well suited	well suited	not well suited	very well suited	not well suited	not well suited	A	very well suited
Sensor Systems Development	not well suited	very well suited	less well suited	very well suited	not well suited	less well suited	A	very well suited
Agent Transport Analysis	well suited	very well suited	not well suited	less well suited	less well suited	well suited	A	less well suited
3. Individual & Collective Protection								
Controlled Molecular Transport Materials Discovery	very well suited	less well suited	less well suited	very well suited	less well suited	less well suited	A	well suited
Barrier Materials Engineering	well suited	less well suited	well suited	well suited	not well suited	not well suited	A	well suited
Personal Protective Systems Development	not well suited	not well suited	well suited	less well suited	not well suited	not well suited	A	less well suited
Collective Protective Systems Development	not well suited	not well suited	well suited	less well suited	not well suited	not well suited	A	less well suited
Physiology	well suited	well suited	well suited	not well suited	not well suited	not well suited	A	less well suited
4. Medical Countermeasures								
Target Discovery	very well suited	less well suited	less well suited	well suited	less well suited	not well suited	less well suited	well suited
Regulatory Science	less well suited	well suited	less well suited	well suited	less well suited	well suited	less well suited	very well suited
Mechanisms of Delivery & Delivery Systems	less well suited	not well suited	well suited	well suited	not well suited	not well suited	not well suited	very well suited
Animal Models	very well suited	less well suited	very well suited	well suited	N/A			
Host Response	very well suited	less well suited	well suited	less well suited	less well suited	not well suited	less well suited	well suited
Pre-Clinical Studies	N/A				well suited	well suited	well suited	very well suited
Clinical Trials (GLP and GMP)	very well suited	less well suited	less well suited	very well suited	very well suited	not well suited	not well suited	very well suited
Medical Product Development	less well suited	less well suited	less well suited	very well suited	less well suited	not well suited	not well suited	very well suited
5. Hazard Assessment, Mgmt and Decon								
Decontamination Methods Discovery	very well suited	well suited	very well suited	less well suited	less well suited	less well suited	A	less well suited
Decontaminant Development	less well suited	well suited	well suited	very well suited	not well suited	less well suited	A	very well suited
Decontamination Resilient Materials Development	less well suited	less well suited	less well suited	less well suited	less well suited	less well suited	A	less well suited
Decontamination Systems Engineering	not well suited	less well suited	very well suited	very well suited	not well suited	not well suited	A	very well suited
Agent Transport & Viability Analysis	well suited	well suited	well suited	well suited	not well suited	well suited	A	less well suited

6. Cross Cutting Science and Technology
Acquisition, Maintenance and Transport of Critical Chemical and Biological Reagents
Simulation
Informatics
Forensics
Education and Training
Behavioral Analysis
Systems Analysis and Engineering
Repurposing Commercial Technologies
Systems Biology
Synthetic Chemistry and Biology
Materials Science
Statistical Measurement Design
Test & Evaluation

In the committee's view, this category is:
- not well suited
- less well suited
- well suited
- very well suited

to provide this capability to CBDP

R&D: research and development
T&E: test and evaluation
OGF: other govenment facilities
A: the actual chemical or biological agent is important for T&E
N/A: the capability does not have a major R&D or T&E component

particularly true for the biological and chemical sciences which lie at the nexus of the S&T requirements of the CBDP.

Finding 3.2: The military laboratory community is not as strongly partnered with key external research institutions and programs as it could and should be. As the United States has a robust S&T sector, the CBDP can and does engage with individuals and organizations external to DoD and the US government, but this typically occurs at the individual project or principal investigator level, and not necessarily on a sustained basis. The CBDP has not systematically promoted institutional ties with academic, industrial (especially pharmaceutical companies), and other non-DoD laboratories or related federal programs.

Recommendation 3.1: The Director, JSTO-CBD, should ensure that the development of a Culture of Collaboration is a high priority for all elements of the chemical and biological defense enterprise. Although information control requirements and contracting concerns have been stated as barriers on both sides to collaboration, these are issues that can and should be addressed. To ensure that the program delivers products based on the best S&T available, the CBDP needs to find ways to partner with the broader scientific community and other federal agencies in areas relevant to chemical and biological defense.

Tech Watch and Adopt

Finding 3.3: There is the potential to significantly improve chemical and biological defense capabilities by using existing technology. Despite the nation's superb biomedical research establishment and the explosive growth of biological and biomedical science that is relevant to DoD as well as the public health community, relatively little of this broad competency has been applied to problems relevant to chemical and biological defense.

Recommendation 3.2: The DASD(CBD) should establish an effective "tech watch and adopt" component within the CBDP to bring innovative solutions to ongoing needs. Program managers and scientists within the CBDP should recognize the importance of technology watch and adoption before a major new RDT&E investment is made. The incorporation of a "tech watch and adopt" concept would have at least the following three elements: (1) mechanisms for searching and identifying relevant breakthroughs in the literature and from the private sector; (2) mechanisms and processes in place for incorporating innovation into the ongoing program for the capability needed; and (3) processes for rapid adoption of "tweaks" that would significantly improve existing capabilities. An adjunct objective would be to get the external performers interested in CBD problems such that they might be recruited to work on the problem.

Linking R&D Community to Operators

Finding 3.4: Separation of S&T performers from the end user is impeding their ability to meet the user's needs. Individuals in the military laboratories noted that understanding more fully the context of their work could assist S&T personnel in developing operationally relevant products, identifying variables or factors that would otherwise be overlooked, and possibly shortening development time. In addition, a stronger relationship between operators and R&D performers could support innovation by enabling informed, collaborative "blue sky thinking."

Recommendation 3.3: The DASD(CBD) should survey the military laboratories and associated facilities to identify strong relationships between S&T performers and the warfighters, and support replication of such interactions across the program.

Simulants for Test and Evaluation

Finding 3.5: Broadly speaking, the *capacity* for test and evaluation to support the needs of the CBDP exists within DoD. Test and evaluation is a core component of the program and important to maintain within DoD at a high level of competency and responsiveness.

Finding 3.6: Much of the current T&E is based on *unrealistic* expectations of how the material or equipment being tested would *actually* be used. The threat, although long-standing, is uncertain. In addition, the lack of connection with the military operators often leads to the omission of realistic simulation of deployment and use environments.

Recommendation 3.4: Because of the economic, logistical, and environmental concerns with actual agent testing, DASD(CBD) should give priority to the active development and production of realistic and relevant threat agent simulants for both outdoor and large-chamber tests. A single simulant, especially for chemical agents, is unlikely to possess all of the same physical, chemical, and/or transport properties of an actual agent; therefore, multiple simulants may be required to fully stress critical design parameters during T&E.

Review of Test and Evaluation Plans

Finding 3.7: Test and evaluation plans apparently are not subject to independent external review. These plans are created internally, and the committee observed little evidence of the use of external expertise to review testing plans.

Recommendation 3.5: For CBD products to be viable for fielding, the Deputy Under Secretary of the Army for Test and Evaluation should require that (1) T&E activities be based on testing protocols that accurately emulate actual operating environments (both threat properties and operator employment) and (2) independent reviews of testing protocols be conducted.

4

Proposed Approach

CAPABILITIES-BASED STRATEGIC PLANNING FOR CHEMICAL AND BIOLOGICAL DEFENSE

As mentioned in Chapter 1, the Chemical and Biological Defense Program (CBDP) mission statement is overly broad and as a result the program appears not to be driven by strategy and planning. It is overly focused on resourcing, and is short on discipline in evaluating the execution. In this chapter, a strategic framework for the CBDP in support of operational capabilities-based planning is described. This approach, combined with some of the programmatic and laboratory-level considerations described in Chapter 5, is intended to provide guidance for coordination and development of consensus within the CBDP community.

The 2004 Joint Defense Capabilities Study on "Improving DoD Strategic Planning, Resourcing, and Execution to Satisfy Joint Capabilities" describes a management approach with increased emphasis on strategy, planning, and accountability (see Figure 4.1). Key elements of this approach are enhanced planning, and execution accountability.

The challenges to the Department of Defense (DoD) in the realm of chemical and biological defense are complex. The Department has responsibilities that span the missions of protecting the warfighter, providing support to the warfighter, defending the United States from attack (i.e., Homeland Defense), and supporting local authorities in executing disaster response following a chemical- or biological-related incident (i.e.,

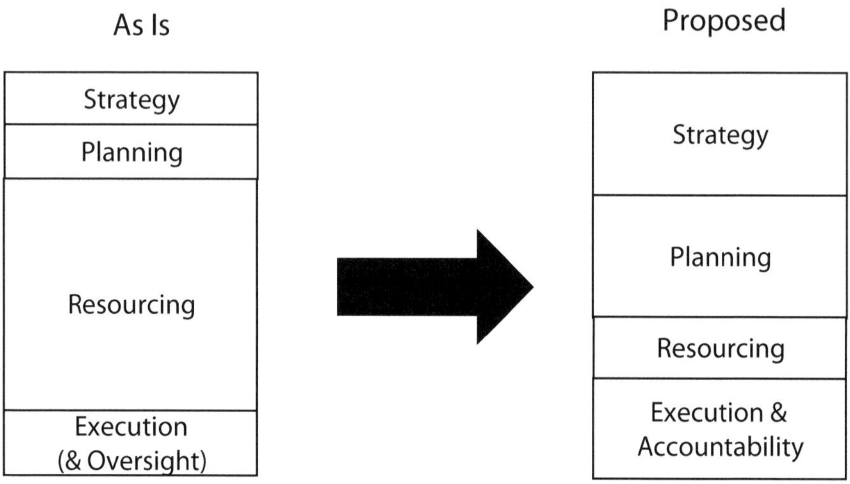

FIGURE 4.1 Notional diagram of the current and proposed management approaches. Box size indicates the relative importance of the element within the approach.

defense support to civil authorities). Events requiring DoD to perform each of these missions could unfold in innumerable, unexpected ways:

- Threats of intentional attack may be unforeseeable.
- Incidents of naturally occurring disease or unintentional chemical exposures cannot be anticipated.
- Where and when the events will occur is largely unknowable.
- Intelligence activities could provide warning of events, but cannot be taken as infallible.
- Adversaries may adopt tactics to counter attempts to defend against attacks.
- Unanticipated events could diminish defense and response capabilities.

The implication of these factors, when considered together, is that it is impossible to describe a concise set of most likely scenarios for which DoD needs to be prepared. In a fiscal environment that demands choices be made among which capabilities DoD can develop and sustain, decision making is even more challenging.

In these contexts, planning often relies on requirements-based processes which describe preferences for individual capabilities based on assumptions of the most likely conditions for which they will be needed.

Priority	Capability
1	Chemical Standoff Detection
2	Chemical Point Detection
3	Biological Point Detection
4	Biological Standoff Detection
5	Respiratory and Ocular Protection
6	Biological Prophylaxis
7	Field Analytics
8	Personnel Contamination Mitigation
9	Integrated Early Warning
10	Radiological Standoff Detection
11	Radiological Point Detection
12	CBRN Reconnaissance
13	Equipment Contamination Mitigation
14	Chemical Prophylaxis
15	Medical Surveillance
16	Percutaneous Protection
17	Medical Diagnostics
18	Battle or Operating Environment Analysis
19	Biological Therapeutics
20	Chemical Therapeutics
21	Battle or Operating Environment Management System
22	Expeditionary Collective Protection
23	Radiological Prophylaxis
24	Fixed Site Contamination Mitigation
25	Radiological Therapeutics
26	Fixed Site Collective Protection
27	Methods of control
28	Remains Disposition
29	Hazardous Waste Control

FIGURE 4.2 Joint Priority List (JPL) from 2011.

These preferences are then translated into ranked lists of priorities (for example, see Figure 4.2, 2011 Joint Priority List). The lists are then used to inform budget decisions via the Program Objective Memorandum (POM) process, with the idea of directing resources toward those capabilities that are higher on the priority lists. This type of approach fails to account for the reality that the scenarios upon which the priorities are predicated are most likely not the events that will unfold and that overall performance depends on interdependencies between the capabilities being developed. In addition, the rigidities of the POM cycle often make the timing of various research, development, test, and evaluation stages critical to project "success," and are not flexible.

An alternative approach is to use capabilities-based planning (see Figure 4.3). Here, the goal is to adopt strategies that are *flexible* enough to provide capabilities for events other than those anticipated, *adaptive* to conditions other than those that are planned, and *robust* to attempts made to diminish these capabilities. Framing decision making in this way de-emphasizes prioritization and optimization of capabilities. Instead, this framing promotes making choices among portfolios of capabilities that balance tradeoffs among mission performance, risks, and costs. The output of this process approach is guidance on which capabilities to pursue.

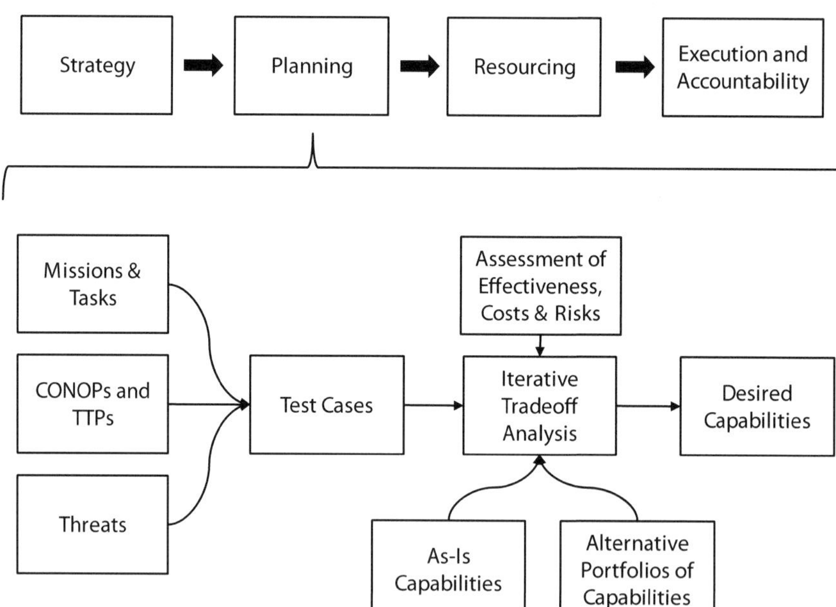

FIGURE 4.3 Diagram of an approach for planning in a capabilities-based process.

This approach has proven useful in cases of deep uncertainty and fiscal constraints in areas such as planning for capabilities related to missile defense and global strike. Capabilities-based planning is hard and can only be undertaken if the resources, expertise, and will are available. When done correctly, it is a powerful approach, but if done poorly, it will lead to confusion and result in new gaps in the program. It should be noted that if the capabilities-based approach is adopted, there may be elements of the current program that should be transferred.

Components of the new approach include identifying a meaningful set of test cases, selecting and assessing sound measures of effectiveness, building creative portfolios of capabilities, and developing tools to conduct iterative tradeoff analysis.[1]

Identifying Meaningful Test Cases

Proliferation of missions, ambiguity about threats, and multiplicity of CONOPS can lead to innumerable potential scenarios against which program portfolios can be assessed. Practicality demands that assessment be constrained to a concise set of test cases. Capabilities-based planning addresses this challenge in two ways. First, the cases used for analysis are selected not because of belief that they are inherently more likely or more important than other possible scenarios. Instead, they are selected based on a view that an option that performs well in the conditions specified in the case will exhibit a capability deemed important—i.e., the case represents a test. Second, those test cases are selected based on the same type of deliberation among analysis communities that is required to build creative portfolios. Striking a balance between relevant and not overly constrained test cases is obviously difficult and requires iteration during the analysis. In selecting the test cases, it is also important to consider the findings of relevant intelligence and threat assessments. Test case development should include red-teaming. Red-teaming should help ensure that casualties are not the sole measure of risk, and that asymmetric and terrorist threats are sufficiently considered.

Selecting and Assessing Sound Measures of Effectiveness

Sound measures of effectiveness are grounded in a clear logic of how mission success is defined and how capabilities are combined to achieve

---

[1] Joint Defense Capabilities Study, *Improving DoD Strategic Planning, Resourcing, and Execution to Satisfy Joint Capabilities*: Final Report, 2004; Davis, Paul K., *Lessons from RAND's Work on Planning Under Uncertainty for National Security*. Santa Monica, CA: RAND Corporation, 2012.

mission success. When tied to such logic models, measures are more likely to be valid and less likely to promote perverse or unintended decisions. Sound measures should also be reliably measureable, particularly when linked to program evaluation. Only then can estimates of the measures for different programs and at different times be trusted. Additionally, to be useful, measurement must be feasible given time and resources consistent with the decisions they are being used to effect. Red-teaming is integral to completing a valid assessment of effectiveness.

Building Creative Portfolios of Capabilities

Policy makers can only expect good outcomes if they have options that include opportunities to balance across performance and costs tradeoffs. Options that are optimized for a specific scenario or capability are unlikely to be flexible, adaptive, and robust. However, developing creative portfolios of alternatives requires iterative deliberation between the warfighter and support operations, science and engineering, systems analysis, and cost analysis communities. Incorporating this deliberation into strategic planning is critical to sound analysis.

Developing Tools to Support Tradeoff Analysis

Capabilities-based analysis requires many types of tools. The multiplicity of test cases requires tools that can allow exploration of performance across a large number of conditions and can be reconfigured quickly to be used for other cases throughout the iterative analytic process. These assessments must be grounded in valid estimates of performance costs and risks. These estimates could come from many sources including red-teaming, modeling and simulation, field demonstrations, and reliably conducted expert elicitation. Finally, tools are needed to illustrate the tradeoffs inherent in choices among alternative program portfolios. The choices supported by this analysis provide guidance for desired capabilities and are the starting point for identifying capability gaps.

Once the desired capabilities are identified, the execution and accountability stage can be started (see Figure 4.4).

Deriving Proposed Solutions and Specifications

After supporting analysis tools have been applied and desired capabilities have been identified, the acquisition community is in a position to propose solutions and develop specifications for those solutions. To be effective, this process should incorporate realistic red-teaming and deliberation between the user and science and technology (S&T) communities

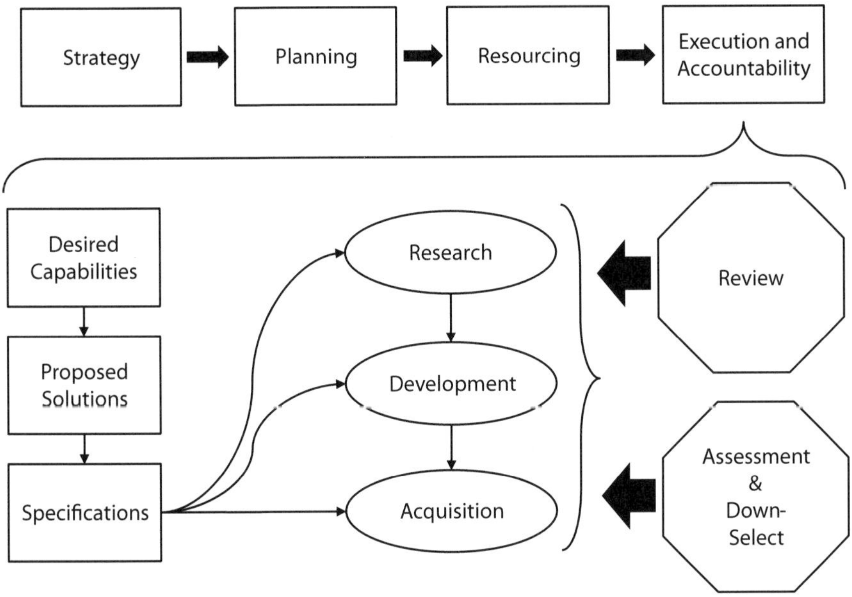

FIGURE 4.4 Diagram of execution and accountability in a capabilities-based process.

so that the process considers both innovation and technical feasibility—the art of the possible and the art of the probable (see "Maintaining a Connection to the End User"). Depending on the urgency, difficulty, and capability base, decisions should be made about the degree of specification needed before a research-development-acquisition (R-D-A) process begins. Specifications should consider whether or not 100% survivability is needed or possible.

From Specification to the Research-Development-Acquisition Process

The R-D-A process is well established within DoD, and many elements of the established process are adequate for the CBDP. For the R-D-A process to be effective for the medical countermeasures program, however, R-D-A should be done as a team approach with end-to-end involvement, including regulatory processes considered in the earliest phases.

Once specifications are derived based on a solid analysis of capability gaps and tradeoffs, then the maturity of existing products can be assessed against the specification to determine whether new, innovation research is necessary (Research) or whether development or furthering of an existing idea is appropriate (Development), or if a developed product is ready and simply needs to be acquired (Acquisition). It may be the case that activities could be started at more than one R-D-A level to build in a need to address near-term needs with a spiral development process that will fundamentally change the product in the future. It would be expected that more projects would be started in the research phase than the development phase, and even less in the acquisition phase. If the R-D-A process is conducted so there are multiple projects and available options ("shots on goal") then it is essential that a robust, independent down-selection process is established. In the development and acquisition phases, regular assessments are also essential. These assessments should evaluate technical quality as well as progress toward project and program goals. Such assessments provide the most credible way to make down-selection decisions.

FINDINGS AND RECOMMENDATIONS

In considering the strategic planning process necessary to support the Chemical and Biological Defense Program, the committee identified the following principle findings and recommendations.

Capabilities-Based Planning, Development, and Acquisition

Finding 4.1: A requirements-driven S&T process is not a good match for the CBDP. The planning and experimentation carried out by the CBDP is usually so removed from plausible use that it is difficult to believe that the Combatant Commands would know how to understand and evaluate the program's impact, how best to protect their forces, to carry out their operations in the face of current and/or high-probability future threats. Planning tends to focus on narrow conceptions of threats and responses derived from historical events. Outcomes tend to be described in terms of consequences which can be easily measured, such as fatalities and injuries. Options tend to be developed based on incremental modifications to current materiel and operations. Each of these approaches is inadequate for addressing the evolving and innovative nature of chemical and biological threats. Moreover, the perceived goal of "100% protection" appears to impact all aspects of the program such that few products reach the field in a timely manner, especially in the medical countermeasures part of the program.

Recommendation 4.1: The Office of the Secretary of Defense (through the Assistant Secretary of Defense for Nuclear, Chemical, and Biological Defense Programs) should evaluate a shift to capabilities-based

planning, as a more appropriate approach for this program. The goal is to adopt strategies that are *flexible* to provide capabilities for events other than those anticipated, *adaptive* to conditions other than those that are planned, and *robust* to attempts made to diminish these capabilities. Planning should expand the range of options considered; iterative review and realistic red-teaming should challenge assumptions built into plans and promote innovations in defense to correspond to that in the threats. The scope of red-teaming and review should encompass the threats and activities against which performance is assessed and the evaluations of performance are made. The overall S&T focus should shift from "zero casualties" to "mission success."

5

Management of Science and Technology at CBDP

Chapter 4 describes a process for capabilities-based strategic planning, but such a shift ought not be implemented without considering programmatic changes that could support its adoption and successful implementation. This chapter highlights two possible areas, the relationship between the end user and the RDT&E performers, and technology transition within the CBDP, where changes could have a significant effect on improving programmatic efficiency, especially if a capabilities-based planning approach were developed.

In addition, since one of the objectives for this study is to provide input that will assist the DASD(CBD) in "identifying the current model for sustaining lab infrastructure within DoD and determining if this is the most effective method for maintaining DoD critical infrastructure," this chapter also presents observations regarding laboratory management within the CBDP.

As emphasized in Chapter 3 and Recommendation 3.1, a comprehensive, effective RDT&E program must support a Culture of Collaboration. Such a culture encourages individuals with expertise across disciplines to share information and effort in order to achieve their goals and enable programs to effectively manage resources, minimize duplication of effort, and identify opportunities for sharing information and facilities. The sections in this chapter largely relate to aspects of the CBDP that can influence the culture of collaboration at the research and program management levels.

MAINTAINING A CONNECTION TO THE END USER

The goal of the CBDP is to put effective tools in the hands of end users to minimize risk from chemical and biological threats. As described in Chapters 2 and 4, there are four identified missions for the CBDP, and the end users may differ between them. The warfighters and those who support them are critical, but, for example, civil defense authorities and personnel may need to be considered in some cases. In the current programmatic structure, CBD priorities are identified by those with operational knowledge, and products to address those priorities are developed by those with science and engineering expertise. Between these two levels, the S&T efforts required to develop the identified tools are recognized and broken into projects, which are then pursued by individuals or small teams, and the connection to the operational context is often lost until late-stage T&E. In the committee's view, this distance between the R&D performers and the end users represents a lost opportunity to allow for multidisciplinary (and multiperspective) consideration of the challenges of CBD in a field or combat situation. In discussion with the committee, R&D performers who had contact with the warfighters directly, whether as a result of collaboration on specific projects (often noted in connection with USSOCOM) or in the pursuit of operationally relevant data in support of development of models, noted great value from interactions with operators. Other individuals in the basic research realm noted that understanding more fully the context in which their work would be used could assist in the development of operationally relevant projects and help identify overlooked variables or factors in their work. Strengthening the relationship between the warfighter and the R&D performers could support the development of specific capabilities by creating opportunities for innovation by allowing for informed "blue sky thinking" between the two groups.

The committee noted that laboratory personnel from some facilities have relationships with warfighters as a strong component of their research and development programs. We suggest that the CBDP survey the facilities to identify where positive relationships exist, between Special Forces or the Services broadly, and seek to replicate such interactions. At the program management level, it may be difficult to encourage the strengthening of such relationships unless and until a capabilities-based approach is adopted as it may be challenging to see how providing R&D performers with greater operational context can be relevant to meeting specific requirements.

TECHNOLOGY TRANSITION WITHIN CBDP

When a project progresses from the initial research and development work performed in support of product development to the test and evaluation and approval and acquisition of the material, management shifts from JSTO-CBD, under DTRA, to JPEO-CBD. Though both offices are overseen by the CBDP, there is no one office or individual who oversees the entire process for any given product. The discussion below highlights some of the challenges that this structure can pose for efficient development and presents some possible alternative options for consideration by DASD(CBD). These changes could help to improve the current processes, should a requirements-driven process be maintained, or support a revised, capabilities-driven process, as discussed in Chapter 4.

The development of a final product is often iterative in nature, with failures in T&E identifying weaknesses that can only be resolved with additional R&D. At such times, the project management shifts from JPEO-CBD, which manages the T&E process, back to JSTO-CBD for additional R&D before returning to JPEO-CBD for the next round of testing and, potentially, approval. Successful transition between these offices requires a shared understanding of appropriate transition points. If an agreement is not in place, there is a risk that a gap between the perceived roles could result in the development of a project management "valley of death" and reduced program efficiency. While such issues can potentially be resolved through good working relationships between the offices, reliance on that relationship for success is not recommended as it can be strongly influenced by personality rather than policy. To address this concern, the committee recommends that DASD(CBD) consider alternative program management methods, including incorporation of an end-to-end project management authority to be maintained by an individual with scientific and technical expertise, as opposed to an acquisition specialist. In addition, liaisons such as those used between the management offices and the COCOMs, or secondments between offices and laboratories could support the development of both formal and informal relationships between program personnel and facilitate the development of trust and understanding between the various parties and offices.

One particularly stark example of the importance of an end-to-end project management authority, or at least consensus on the appropriate transition point, is in the development of medical countermeasures (see Box 5.1 highlighting some failures of medical products where poor transition between development stages was a contributing factor). The root of the difficulty is that development and acquisition of these items must contend with regulatory requirements—e.g., good laboratory and manufacturing processes, clinical trials, and acquisition of safety and efficacy data—and these requirements must be met in order to obtain FDA

BOX 5.1
Examples of How Medical Product Development and Fielding Has Failed

The transition from basic research to advanced development for medical CB countermeasures has been difficult and slow. The traditional approach in DoD's medical laboratories has been to do the initial research, identify and characterize candidate products, perform pre-clinical testing, and prepare some sections of an Investigational New Drug (IND) filing. The subsequent steps that are necessary to move such products through the rest of the acquisition process and the FDA regulatory process have been stymied for a variety of reasons and can be typified by these examples:

1. rPA vaccine: DoD conducted necessary S&T to identify and characterize Protective Antigen (PA) as the primary antigen for a new anthrax vaccine. Using recombinant DNA technology, GMP rPA was produced in pilot laboratories, safety and efficacy was shown in appropriate animal models (up to and including non-human primates) and the IND application was filed with the FDA. Unfortunantly, no practical expertise in obtaining scale-up development capabilities was resident in DoD; HHS took responsibility for production of large-scale lots of vaccine, clinical trials, and bringing the candidate though the FDA process. Technical issues related to the product stability, lot-to-lot variability, and other issues could not be resolved by DoD or HHS. rPA has yet to become a licensed product.
2. RSDL (Reactive Skin Decontamination Lotion): The extraordinarily long process to achieve FDA approval of this product can be attributed to the fact that DoD laboratories did not have sufficient expertise to address some of the regulatory issues related to product safety, incompatibility, and/or efficacy. Collaboration with experts from the drug development industry could have addressed some of these issues. RSDL was cleared by FDA for military use in 2003.

approval of the final product. As these requirements necessarily influence product development pathways at an early stage, the current management structure within the CBDP is not well suited to the task because of the lack of a whole-process, integrated view of product development. In addition to challenges presented by the existing program management structure, the medical production process has not traditionally been a major focus of DoD, and as a result, the in-house expertise at the decision-making process is likely to be limited.

Medical Product Development within the CBDP

Development of therapeutics under normal operating conditions within the pharmaceutical industry is a long and expensive process (commonly viewed as requiring "twelve years and one billion dollars"). Being successful necessitates extensive management and quality control features. The complexity is such that a team management approach, drawing upon internal expertise in the early development phase through to the later production phase, is a common approach for incorporating continuity through the entire process. Development of therapeutics in the context of CB agent exposure adds complexity to the process:

- Documentation of pathophysiology of the disease in humans is minimal for most of the targets, and the efficacy of the medical products cannot be tested in humans against the relevant pathogens. Thus, extrapolation from efficacy studies in animals are required to predict efficacy in humans;
- Animal models are difficult and not well defined for many of the pathogens;
- The regulatory process for developing therapeutics, particularly vaccines, is not a well-delineated process in particular with relationship to the requirements for correlates of protection;
- The acceptable risk for the therapeutic efficacy range appears to be set at a very high threshold, well above what is typically acceptable under other military operations.

The reason for listing some of these complexities is to demonstrate that production of vaccines and therapeutics in the DoD CBD context is extraordinarily challenging, and careful planning and project management is required to be successful. Over the course of the committee's data-gathering, it was apparent that the current process for development of medical products has led to few, if any, new therapeutics. Discussions with CBDP staff indicated that the following factors are contributing to the low success rate:

- Minimal communication between the JSTO-CBD and JPEO-CBD managers to facilitate product development;
- An apparent expertise gap in the management offices, particularly regarding identification and handling of critical transition points in product development;
- Turnover in personnel at several levels without a committed legacy of ongoing product development makes for multiple "starts and stops" during the process;

- Lack of a transparent process for determining long-range critical therapeutic or vaccine targets; and
- No consistent methodology for selection of partner laboratories or companies with expertise in therapeutic development.

Note that these factors were not identified as part of a formal review of the program, but through formal and informal conversations. The committee believes that a formal, preferably external, review of the process would be valuable before taking action to modify the current process.

Establishing "Common Language" for Transition of Medical Countermeasures

One possible model for managing medical product development transitions is demonstrated by the integrated portfolio for CBRN countermeasure development through BARDA. BARDA's role is to develop medical countermeasures for the Strategic National Stockpile, and as a result, it shares many of the challenges DoD faces in its therapeutic development process. Though housed within HHS, BARDA represents a collaborative effort between multiple agencies, including FDA, DoD, and DHS.

One of BARDA's policies is to not accept potential products into the program until the work has progressed through an FDA-approved Phase I safety study. Using this structure to define the benchmark allows for clear communication of expectations across various agencies and companies. Using an FDA regulatory step as the transition point from JSTO-CBD to JPEO-CBD—or to or from another DoD/USG development partner—could provide a similar "common language" determining the transition point and establishing expectations within the CBDP. Such an objective transition point may improve project and personal efficiencies by allowing for

- more accurate out-year staff and budgetary planning, which would hold value not only for the project scientists and managers, but also for higher-level commanders and directors for portfolio and/or budget planning and defense;
- reduced burden of negotiation meetings to establish transition points for ongoing projects;
- clearer determination of the responsibilities and expectations of JSTO-CBD and JPEO-CBD; and
- easier to manage timelines for expected returns on R&D investments.

BOX 5.2
Defining a Transition "Trigger"

DoD is not the only agency or organization that faces the challenges described herein. Others, such as pharmaceutical companies in the commercial sector and BARDA within the US government also must manage transitions effectively. There are three commonly used and referenced transition points early in a product's development that are seen as logical "triggers" for the transition from basic research to development of a product. They are (1) submission of an investigational new drug (IND) and (2) completion of the Phase I trial of a material.[a] As one example of how these triggers could be incorporated into the CBDP, the Material Development Decision (MDD), which currently resides with JPEO-CBD, could become trigger for initiation and development of an IND application for submission to FDA. Practically, this would mean all discovery and preclinical activity would reside under JSTO-CBD management, and successful programs would be presented to JPEO-CBD for a MDD. MDD approval would trigger construction and submission of the necessary applications to enter an FDA-approved regulatory approval path. Alternatively, transition to JPEO-CBD management could occur after completion of a Phase I trial.

Under a third alternative (3), as part of a broader strategy within the CBDP for FDA-regulated products to more efficiently use available advanced development funds, successful programs could be "parked" after construction of the IND or after Phase I trial completion. This could be especially useful when the program has multiple potential products in any given area of need. Regardless of the chosen trigger, expertise, within or contracted by JSTO-CBD and JPEO-CBD, needs to be appropriately positioned. This approach would also be supportive of overlap in JSTO-CBD and JPEO-CBD personnel engagement on the project to ensure smooth and knowledgeable transitions.

---

[a] Selection of either an IND submission or Phase I completion in the common JSTO-JPEO transition point would generally align with DoD 5000 service recognition milestones A ("first in human") or B ("point of concept"), respectively. It would also generally align with transitions from 6.2 to 6.3 or 6.3 to 6.4 funding.

An additional potential benefit would be the inclusion of FDA scientists, such as those within the Medical Countermeasures Initiative (MCMI) program. Clearly, the earliest possible settings of expectations from the FDA—even from an advisory position—on the types of data sets that may be required to reasonably complete a Phase I trial would be helpful in managing the overall process. See Box 5.2 for a discussion of possible transition points for medical product development.

LABORATORY MANAGEMENT

As was described in Chapter 1, the RDT&E elements within the CBDP draw upon and direct resources within a number of different laboratories and facilities. These include service laboratories, medical laboratories and facilities, and test and evaluation facilities. Each laboratory and facility has its own mission and management structure, and this has implications for the ability of the CBDP to successfully manage its projects and programs (see Box 5.3). This section presents a brief overview of the history of the management of laboratory research at medical and non-medical facilities, identifies elements of laboratory management the committee feels are critical for success, and presents possible alternative methods for

BOX 5.3
The Role of Medical Laboratories in Chemical and Biological Defense Research

In the 1980s and 1990s, the funding at medical facilities for chemical and biological research was provided by DTRA directly to the commanders for their allocation and distribution. At that time, the laboratories performed basic science in support of understanding the medical response to exposure to chemical or biological agents. The research was not focused on the development of specific products, such as vaccines or countermeasures. Beginning in the early 1990s, DTRA began directing these R&D funds through MRMC (US Army Medical Research and Materiel Command), giving MRMC's command structure both responsibility and authority for the science within their major laboratories. Under this new model, funding was allocated by program and, as a result, the laboratories became focused on the development of specific product targets, e.g., an anticonvulsant or a plague vaccine, though still on the basic science in support of that target. Advanced development leading to an FDA-licensed product was not within the laboratories' purview.

When the JPO was formed, the role of the laboratories changed again. Now medical laboratories, specifically USAMRIID and USAMRICD, were asked to take candidate countermeasures beyond the basic science toward the development of a licensed product. This required the laboratories to produce pilot-scale lots of GMP (good manufacturing practices) candidate product and to conduct the GLP (good laboratory practices) pre-clinical studies in preparation for the initial submission to FDA for approval.

This continued until the formation of JSTO-CBD in 2003. Though the focus on specific product development continued after that point, investigators within the laboratories were asked to respond to specific requests for proposals in a competitive environment, rather than the laboratories pursuing a program (e.g., development of an anthrax vaccine) with guaranteed funding. With this change, the responsibility for performance remained with the laboratories and the authority to manage the programs resided at JSTO-CBD in Ft. Belvoir.

laboratory management that could be considered. Overall facility management, which includes infrastructure, workforce, and research program management within the facility, plays a critical role in the stewardship necessary to support a successful S&T related endeavor. This stewardship also extends to funding and directing the scientific work of the program.

Successful Laboratory Management

The CBDP relies upon a laboratory network and test ranges to provide RDT&E and produce products critical to the chemical and biological defense of the nation. The network consists of DoD owned and operated laboratories (e.g., USAMRIID, ECBC, Air Force Research Laboratory (AFRL), NSRDEC, and Naval Surface Warfare Center (NSWC) Dahlgren), DOE national laboratories (e.g., LANL, SNL, LLNL, Pacific Northwestern National Laboratory (PNNL)), FFRDCs (Federally Funded Research and Development Centers), non-profit entities (e.g., Battelle), for-profit commercial laboratories, and universities. At many of the facilities the committee visited, the recent construction of new buildings will provide new capacity for RDT&E, adding both additional space and additional technical functionality.

Successful RDT&E leading to fielded systems[1] is a long (perhaps a decade or more), arduous process. In the committee's view, a successful RDT&E program requires at least the following six elements to ensure clarity of purpose, focus of investments, and coherence of management:

1. Clear mission and objectives
2. Continuity in leadership
3. The ability to understand, accept, and manage risk throughout the process
4. Predictable and stable funding
5. Effective asset management at the laboratory level
6. A sense of excitement and pride in the work among the staff

In the following sections, these elements will be defined and subsequently discussed in the context of the CBDP.

[1] Fielded systems are not exclusively hardware or medical countermeasures. For example, they may include an operational diagnostic and analysis systems including collection, transportation, analysis, and identification including all the required infrastructure, protocols, routinely exercised.

Element 1: Clear Mission and Objectives

For an effective program, all participants in the endeavor need to understand its mission and objectives, how they contribute to achieving those goals, and why their work is important in that context. A laboratory cannot be successful if it proclaims it is undertaking all missions for the entire government. "Reinvention" may be required as the world changes, but this should only happen rarely. Without a clear, defined mission and objective, so-called "mission confusion" may lead to expansion into activities that do not directly support the objectives or that are duplicative of efforts in other laboratories. Lessons from successful laboratories (government and non-government) demonstrate that a shared mission understanding leads to long-term success.

Element 2: Continuity in Leadership

Strong leadership, and continuity of the stated mission and objectives during and after changes in leadership, supports the development of sense of mission within the workforce. In contrast, rapid changes in leadership and/or weak leadership can contribute to the mission confusion described above. Of course, sometimes rapid changes in facility directors or commanders is unavoidable, but in such cases, care should be taken to ensure that succession planning, pre-training, orientation of incoming personnel, and personnel continuity in senior positions is encouraged. These steps will help maintain a consistent vision and understanding of the facility's mission and objectives, which is important for pursuit of programmatic and project-related goals. An additional benefit is that when individuals are in a position for a significant period of time, e.g., longer than two years, they have a greater ability to develop relationships and partnerships within the research and end-user communities than might otherwise be possible. This is beneficial for the identification of opportunities for collaboration.

Element 3: The Ability to Understand, Accept, and Manage Risk

RDT&E is a risky process, and a successful research and development program may have many false starts before achieving success. Thus, an essential element of managing such programs is accepting and balancing risks. There may be times when it is appropriate to undertake high-risk, high-payoff projects which may ultimately fail in addition to maintaining long-term focus on a specific area with low risk in order to provide fundamental understandings that support new developments. An effective program will create a balanced portfolio of these various research types. Similar to the process described in Chapter 4, in the section "From

Specification to the Research-Development-Acquisition Process," external review can help down-select projects to improve overall chances for programmatic innovation and success with reduced overall costs.

Care should also be taken to not introduce unnecessary risk of project failure through poor planning and management. Understanding the eventual goal of a project or program and identifying appropriate milestones that must be completed or addressed for success can allow for corrections of approach and provide confidence at critical junctures that all required elements are in place to minimize the chances of failure due to bureaucracy.

A necessary adjunct to a well-balanced risk approach is a method for continual assessment of program and/or project progress. Internal, and especially external, standing technical review committees (supplemented by those with operational knowledge) are required to ensure that unsuccessful programs are terminated, to provide technical review for high-risk/high-payoff projects, and to encourage consideration of programs that may not have originated within the facility. The committee cautions that each institution might appoint separate review boards with different membership; while a diversity of opinions is good, having numerous, separate groups could prevent identifying redundancy and duplications. It is essential that reviewers are able to consider the context and larger picture, and maintain continuity.[2]

Element 4: Predictable and Stable Funding

R&D takes time. Continual disruptions due to major funding shifts and delays lead at best to inefficiencies and, at worst, failure. Funding can roughly be considered at three levels (project, program, and laboratory):

- **Project.** Projects are narrowly scoped activities with specific, well-defined goals. Projects may only require a few years for completion, but this timeframe places a high premium on efficiency. Gaps in funding can hinder that efficiency, and an uninterrupted funding stream for the duration of the project or at least between major project milestones/decision points is preferred.
- **Program.** Programs have a broader scope than projects with desired outcomes that require multiple steps or inputs for completion. These can last up to a decade, and to be successful, funding needs to be assured (to the maximum extent possible) for that

[2] As one possibility, there are several existing government groups with non-governmental life scientists and others who hold clearances at sufficiently high levels and could be utilized, with augmentation as needed.

duration. Some adjustments in priorities are inevitable, but major directions must be consistent and supported.

- **Laboratory.** Laboratories exist for multiple decades and require predictable funding to ensure a high-quality workforce and up-to-date research facilities and infrastructure. Laboratory directors and commanders must be assured of some level of funding if they are to effectively plan for the future and manage a workforce and research infrastructure. This should not be confused with entitlements, where funding is continued independent of performance. In order to be effective, a balance should be found with significant amount of core funding "guaranteed" and the remainder provided through a competitive environment.

Element 5: Effective Asset Management at the Laboratory Level

Some local control of funding is considered by the committee to be an important component of successful laboratory management. A local ability to move people and resources between projects can assist in creating an efficient environment as decision making can be more responsive than if external approval is required. If a laboratory director is prevented from anticipating and acting on predicted future needs, the laboratory may find it challenging to capitalize on emerging technologies or provide for new operational needs. Workforce management should be left in the hands of laboratory management (with command oversight distinguished from management), with external laboratory reviews providing input to help identify emerging needs or areas of concern (see Element 2). Similarly, programs may benefit when laboratory directors have some amount of unallocated funds (internal research and development (IR&D) and capital funds) to invest as they deem necessary (e.g., providing funds to a researcher to pursue a new line of inquiry or addressing unmet maintenance or operational needs) to ensure that the laboratory will be able to continue to meet its mission obligations. This is part of instituting a balanced risk approach to RDT&E.

Element 6: A Sense of Excitement and Pride in the Work among the Staff

Ultimately the success of an RDT&E program depends on not only the creativity and skill of the staff, but also the environment in which they work. Modern research by its very nature is a collaborative effort. Engagement with the research community, both internal and external, gives research staff access to a diversity of ideas, which may open new areas of research or help make intellectual connections between different projects (see Box 5.4). Networking between facilities and researchers

BOX 5.4
Description of the Human Component of S&T Capability

Professional development of performers, program, and middle and senior management is also important in creating a positive working environment and recruiting and maintaining quality within the program. While the CBDP should seek to recruit and retain the best possible personnel for each position across the enterprise, a coherent system of professional development is also needed. Without entry-level development, new performers or entry-level program managers should be provided indoctrination as to requirements and processes associated with their positions as well as exposure to their customers and stakeholders. From the outset they should have a clear understanding of why their program exists and who it serves. As experience is obtained and acceptable performance is maintained, broadening opportunities should be provided in relevant science and with organizations and experts who are working in closely related or complementary areas (both inside and outside of DoD and the USG). Regular professional development opportunities should be provided at specified intervals, based on individual qualifications, performance and career goals and the needs, requirements and strategic goals and objectives of the CBDP and subordinate components. Opportunities to refresh in relevant S&T should be part of professional development and will benefit the individual and program. "One size fits all" will not be an appropriate or value-added approach.

can also help identify potentially underused facilities or resources that can be repurposed or exploited by others in need of additional capacity (note that this can also result in identification of opportunities of "Work for Others" funding). Collaboration also provides an opportunity for researchers to be recognized by their peers as leaders or experts within a given area.

Engagement should not be limited to interactions between researchers. It is especially important to form collaborative efforts with the end users as this can sometimes lead to solutions to unacknowledged or unknown operational needs. These connections can also provide needed field feedback to developers of materials and systems. These collaborations can also provide a reinforcement of the overall mission of the agency and purpose of the program in support of Element 1. This particular point, the relationship between the end user and the developer, will be discussed in greater detail later in this chapter.

LABORATORY MANAGEMENT WITHIN CBDP

The committee visited a number of facilities and spoke with many individuals, both formally and informally, over the course of the study. What follows is a broad summation of the observations and impressions of the management of RDT&E in the CBDP in the context of the six elements outlined in the previous section. Examples are provided where possible and appropriate. The committee recognizes that its interactions with the program have been necessarily limited, and these descriptions are not drawn from formal surveys or metrics. Also see Box 5.5.

Clear Mission and Objectives

The official CBDP mission is to "provide global chemical, biological, radiological, and nuclear defense capabilities in support of National Strategies." This is a broad statement and can be applied to a variety of activities. In conversations with the committee members, facility personnel and program-level staff were in accord with this mission. However, the committee heard different responses with regards to the specific role of given offices, facilities, and laboratories in meeting this mission. Specifically, there seemed to be a lack of common vision between JSTO-CBD and the funding recipients. The recipients expressed concern that it was unclear how the requests for proposals and the competitive process fed into the strategic vision of the laboratory and the CBDP because the proposals did not seem to build on each other to build a comprehensive picture but rather reflected the identified, internal needs or concerns of the year. For the laboratories focused on the development of specific, complex products, an unclear mission or program objectives may hinder the ability of the laboratory to meet its goals.

Continuity in Leadership

Leadership plays an important role in communicating mission and objectives, and in S&T research, this role is often performed by a technical director with a long history within the organization and/or field of research. However, at a number of facilities within the CBDP, the facility military commander (or deputy commander) now changes every two years. Such rapid changes in leadership have the potential to disrupt the ability of an organization to meet its mission and objectives. In such cases, the senior civilian leadership is familiar with the program, which provides a level of continuity, but it is probable that longer-term appointments in the military leadership positions could be helpful in maintaining continuity of programmatic focus at facilities.

The Ability to Understand, Accept, and Manage Risk

The committee recognizes that understanding, accepting, and managing risk in the CBDP is a difficult task. Within the RDT&E program, the research ranges from developing a fundamental understanding of the interactions of molecules with surfaces to developing reliable vaccines and medical countermeasures to be given to warfighters in theater. This being the case, careful management of programmatic risk is essential to ensure that effort and resources are directed appropriately, and there are tools that can be used to assist in this process, e.g., linking activities to a clear mission and set of objectives (see previous section) and robust external strategic and technical review processes.

With regards to external technical review, the committee saw little evidence that peer review of S&T is occurring or encouraged. If this is indeed the case, this is a lost opportunity to draw input from knowledgeable individuals who do similar research, to build connections to those researchers with relevant knowledge, and most importantly, to ensure that promising ideas are not lost and that fundamentally flawed projects do not continue any longer than necessary to identify the flaws. One concern that was mentioned to the committee was that innovative ideas have little room to be developed in the current funding environment, which may result in lost opportunities. With regards to strategic reviews, some facilities have engaged in individual efforts to develop a strategic plan, but those that have not done so within the last decade should be encouraged to create one. In addition to facility-level reviews, a strategic review of the CBDP could identify gaps between mission and activities and allow an opportunity for correction of any such issues.

The committee also notes that poor project management can introduce a higher risk of failure to meet an objective than the science might suggest. One area where this can be seen most clearly is development of licensed medical products. Coordination between multiple offices within DoD and with the FDA are required for successful transition of a prototypical medical product to clinical trials and ultimately to licensing. A poor understanding of the complexities of this process and a lack of end-to-end authority that manages the process can introduce risks of bureaucracy causing a failure to meet the objective. Within the CBDP, the committee notes that few medical products have successfully made the transition to licensing, and though many factors lead to such a situation, a review of the structures supporting this process by the DASD(CBD) may be merited.

BOX 5.5
Waging Science: Re-casting Defense Science Management

To an outsider, DoD's management of CBD science to aid the warfighter and the homeland appears to be run as an acquisition management problem, rather than a science management one. Currently, DoD CB science does not provide an efficient and productive path from needs to solutions. Making material improvements to this situation will not be accomplished easily; however, the committee believes that DoD is uniquely positioned to re-cast defense science management into a much more effective "program" by treating it in a fundamentally different way:

- ***Science is a long-term campaign, not an acquisition of beans and bullets.*** Whether it is developing next-generation protective gear, rapid field diagnostic assays, new vaccines, or standoff CB detectors, there are significant R&D hurdles to be faced before the acquisition phase of "off the shelf" items can be reached. Frequently, multidisciplinary teams are required to tackle problems and eventual success in science is never guaranteed.
- ***Meeting science challenges requires a strategy which requires long-term planning; treat scientific challenges like planning for campaign.*** Treating a scientific challenge more like a mission than an inconvenient prelude to an acquisition will provide the mindset needed to focus the right team of scientists towards a single goal.
- ***The scientific program should be directed toward mission goals and to the building of an ability to react to emerging or unexpected threats. These goals should be understood by all involved.*** Examples of clear goals might be "Create effective diagnostics for all major would-be pathogens"; "Create an effective standoff biodetector that will permit sufficient warning time to don personal protective gear." It is important that all scientists and engineers on the mission team understand that the team mission goal is more important than any organization or personal goals. Examples of such goals might include "Author a major paper on would-be pathogens"; "Keep Laboratory X's share of biostandoff funding at #1."
- ***Both strategies and tactics will be important.*** The mission needs to be planned and managed end-to-end by someone with overall mission leadership and accountability. This is different from current DoD science management where "handoffs" are supposed to occur at various Technical Readiness Levels between organizations that may not be well coordinated end to end.
- ***Plan for the unknown and the unexpected; anticipate change but don't get hung up on predictions.*** Long-range planning to address major defense science challenges simply cannot be effectively planned in DoD standard 5-year cycles. Some challenges will prove much harder than expected, while in other cases new technology may make other challenges far easier than initially planned. About the only certainty is that any multiyear plan involving CBD science challenges will require major revisions by the second or third year, if not sooner. The team and its leadership need to have the capability to react to major changes in plans and still achieve victory.

- ***Assemble the right force to meet the challenge; teaming and coordination will be paramount to achieve victory.*** When planning for a major military battle, the commander in charge will attempt to assemble all the different kinds of forces needed to perform the expected tasks. This might involve land, sea, and air assets, plus satellites. Similarly, the commander of a science battle may need to assemble forces from multiple DoD laboratoriess, academia, industry, and other government science resources. Regardless of which service the commander belongs to, the assembled team must be the best available to achieve the mission goal. Other possible parochial goals ("Laboratory X doesn't have the right expertise but they need some more funding this year"; "Laboratory Y has worked in this field for several decades, but they aren't up to speed on the latest technologies") cannot drive team selection. It is possible that for some scientific challenges most or all of the best expertise needed might lie outside DoD laboratories.
- ***Each battle campaign needs a clear overall leadership chain.*** Waging science is not the same as a NIH or a DARPA basic research project. You do not win a battle with a number of autonomous commanders (principal investigators) who may be working toward different personal goals. Major DoD CB science challenges are inherently Big Science projects that need to be managed appropriately. The skill of effective team formation and motivation cannot be overestimated as necessary leadership criteria when it comes to managing Big Science challenges.
- ***Embrace innovation and be flexible enough to realize when it is needed.*** Technology in the CB domains is evolving at warp speed these days. This is especially true in biology, where recent advances in genome sequencing have increased throughput by at least three orders of magnitude in the past three plus years. The management of DoD CBD science campaigns must be capable of understanding such rapid evolution and be able to manage to make major changes in strategy at the right times. This could be achieved with the establishment of an effective central "tech watch" component within the DoD science portfolio, but each campaign leader would still have to perform their own judgment of whether or when a technology jump is optimal for reaching their mission goal. Innovation can't be mandated, and needs to be nurtured. Campaign leaders need to have the ability to have their teams test out promising innovative ideas.
- ***Logistics and infrastructure win wars; have a flexible procurement strategy to keep your troops well-equipped as conditions change.*** The experiences in Iraq and Afghanistan painfully illustrated the high cost in terms of life and limb when a combination of factors delayed getting more effective protection for IEDs to troops. Inefficiencies in DoD procurement also affect every CB science campaign, particularly in cases where flexibility from a set plan is needed to deal with technology evolution.
- ***Set meaningful milestones and adjust them as needed based on how the battle develops. The goal is to win the battle, not follow a static plan.*** Improvements that allow and enable innovative flexibility are sorely needed. Paperwork and inflexible regulations appear to cause major inefficiencies in how DoD wages science. Commanders need to have ways to innovate and react to improve their chances of winning.

Predictable and Stable Funding

The committee was informed that, though the funding mechanism has changed over time, the funding levels for RDT&E have remained relatively stable in recent years. In general, discussions with researchers indicated that necessary resources are available for current activities, though multiple groups noted that any significant cut in funding would result in cuts to programs.

One concern expressed many times to the committee was that the competitive process has resulted in a sense that funding is unpredictable, making long-term endeavors challenging. Once a proposal had been approved, the committee heard no concerns regarding funding levels of a specific project. However, timing of receipt of funds was discussed, and delays in receipt of funds due to contracting difficulties were cited as a barrier to developing collaborations with external organizations. At the program level, the primary concern expressed to the committee was that the links between the project-level funding and the program-level priorities were not always clear. With regards to laboratory-level funding, care should be taken to ensure that the recent construction of new buildings, some of which will place new demands on operations and maintenance budgets, does not cause the resources of laboratories and facilities to become overstretched.

Effective Asset Management at the Laboratory Level

Funding for CBD research at the laboratories is awarded through a competitive process directly to the principal investigators. Barring the funds acquired through reimbursement of services provided to external agencies and entities (up to 2.5% of IR&D), the laboratory directors have little direct control over their budgets and the allocation of those budgets to the staff and facilities. During multiple site visits, the committee heard that though people felt they had the resources required to perform their work, there was a lack of flexibility of funding that could make it difficult to be responsive to emerging opportunities or concerns. Some of these concerns could be exacerbated in the near future with the opening of multiple new facilities with new capabilities and, likely, management challenges.

Some researchers who spoke with the committee expressed some concern regarding the degree of JSTO-CBD's involvement in directing research. While it was acknowledged that a competitive process can assist in maintaining a strong research function, concerns that the focus was too great on projects rather than programs was expressed—i.e., no long-term vision or strategy supported the specific calls for proposals. This, of course, must be balanced against the views from the program managers who stated that the proposal requests are a product of the requirements-development process and are reflective of the needs of the CBDP and the agency as a whole. The role of this committee is not to decide who is "right" in this discussion, but to observe that the relationship between JSTO-CBD and the service laboratories appears strained. The lack of local autonomy (authority) within the current funding model is seen as implying a lack of trust, whether intended or not, and this seems to affect the relationship between the two groups. Whether these issues can be resolved through increased communication about the overall funding strategy or if a major revision of the program will be required to resolve these issues will have to be determined and addressed by the DASD(CBD).

A Sense of Excitement and Pride in the Work among the Staff

In light of the discussions above, as one might expect, significant frustration with the funding mechanisms and priorities were expressed many times to the committee. However, in general, staff at the facilities expressed a pride of purpose and a sense that the work being done at the laboratories is important and relevant.

One area that was highlighted by many PIs and performers was a desire to gain a greater understanding of end-user needs and the environment and circumstances in which the materials, products, and procedures will be used. For some, the connection between the research laboratory and the field is rather abstract. Reducing the gap between the bench and the end user could be beneficial in many ways, including increasing excitement and pride in the work among the staff.

Review and Assessment

All projects benefit from scientific peer review when done well, and these reviews keep the skills of scientists and engineers sharp. Reviews also provide an important function in ensuring that scientists within dedicated institutions stay in touch with the external community. Within DoD CBD laboratories, there was a noticeable lack of connectivity to research in other institutions and an independent review process, especially in the critical research phase.

FINDINGS AND RECOMMENDATIONS

This final chapter has brought forth a series of principle findings and recommendations that can assist the program in operating a successful laboratory environment.

Program Management

Finding 5.1: Successful transition between the JSTO-CBD and the JPEO-CBD offices requires a mutual agreement on appropriate transition points, encoded in multiyear program plans and budgets. Regardless of the chosen trigger, expertise and resources within or contracted by JSTO-CBD and JPEO-CBD need to be appropriately positioned. This approach would also be supportive of overlap in JSTO-CBD and JPEO-CBD personnel engagement on the project to ensure smooth and knowledgeable transitions. However, the committee observed that the partnership between the JSTO-CBD and JPEO-CBD is weak and that neither office viewed transition plans as a responsibility.

Finding 5.2: There is no end-to-end authority for the CBDP, which is particularly problematic for medical products. Though both JSTO-CBD and JPEO-CBD are overseen by the CBDP, there is no one office or individual with the responsibility and authority for the entire process for any given product. The risk—and reality—is that a transition gap between R&D and acquisition could result in the development of a project management "valley of death." The existing research-development-acquisition process may be adequate for acquiring the non-medical products in the CBDP. For the medical countermeasures program, however, FDA regulatory requirements must be considered early enough to influence product development decisions. The current management structure within the CBDP is not well suited to the task because of the lack of a whole-process, integrated view of product development.

Recommendation 5.1: The DASD(CBD) should evaluate alternative program management approaches, including incorporation of an end-to-end project management authority, especially for the medical countermeasures program.

Laboratory and Major Facility Management

Finding 5.3: The principal RDT&E military organizations associated with the CBDP are benefiting from major facility investments that are planned to provide both capabilities and capacities to meet the anticipated needs of the program. Operating and maintaining these facilities, however, will place a burden on both the owning Service (principally the Army) and the program. The initial operating plans appear to be resourced.

Finding 5.4: All or part of the elements required for healthy RDT&E activities were missing at the organizations visited by the commit-

tee. A successful RDT&E enterprise should include the following elements to ensure clarity of purpose, focus of investments, and coherence of management:

1. Clear mission and objectives
2. Continuity in leadership
3. The ability to understand, accept, and manage risk throughout the process
4. Predictable and stable funding
5. Effective asset management at the laboratory level
6. A sense of excitement and pride in the work among the staff

Of special concern are strained relationships between JSTO-CBD and the laboratories, the new rotational policy for military commanders in the Army, and a trend toward increasing oversight of both technical work and operations at the facilities.

Recommendation 5.2: The DASD(CBD) should formally review alternative laboratory management models, taking advantage of the numerous prior studies, reviews, and evaluations of laboratory and large facility management of S&T organizations. A principal objective is to define the level of stewardship that the program should provide to the principal RDT&E in-house facilities and laboratories.

Scientific Peer Review

Finding 5.5: All programs benefit from scientific peer review when done well, and these reviews keep the skills of scientists and engineers sharp.

Recommendation 5.3: The DASD(CBD) should implement a nested review process for chemical and biological defense RDT&E bound by consistent standards of rigor, frequency, and reporting. The CBDP and its supporting laboratories would each benefit from independent, periodic review at the programmatic and scientific levels. The CBDP should also encourage and participate in institutional reviews. An annual roll-up of review outcomes could help identify thematic areas of promise and concern.

Appendix A

Schedule of Data-Gathering

Data Gathering Meeting 1:
February 29-March 1, 2012 in Washington, DC

The committee received overview presentations of the CBDP from the following people:

Dr. Gerald Parker, Deputy Assistant Secretary of Defense for Chemical and Biological Defense
Dr. Robert Cohn, Chief Scientist, Office of the Assistant Secretary of Defense for Nuclear, Chemical, and Biological Defense Programs/Chemical and Biological Defense
BG Lucas Polakowski, Deputy Director for Force Protection and Counter Weapons of Mass Destruction, J8
BG Jess Scarborough, Joint Program Executive Officer for Chemical and Biological Defense
Mr. James Cooke, Deputy Under Secretary of the Army for Test and Evaluation
Mr. Lenny Izzo, Joint Requirements Office for Chemical, Biological, Radiological, and Nuclear Defense
Dr. Alan Rudolph, Director, Chemical and Biological Technologies Directorate, Defense Threat Reduction Agency (DTRA)

In addition to these individuals, the committee heard panel presentations focused on specific capability focus areas. Participants in these panels came from multiple offices from throughout the CBDP enterprise. The focus areas were Surveillance, Medical Countermeasures, Enabling Technologies, and Hazard Mitigation. The committee also received a threat assessment briefing.

Data Gathering Meeting 2:
March 15, 2012 in Dugway, UT

A subgroup of the committee visited Dugway Proving Ground in Dugway, UT. The group met with representatives of the facility and heard overview presentations about the site's capabilities and activities. The group also held roundtable discussions with local project managers, senior scientists, and management personnel. The committee toured the Materiel Test Facility, the Combined Chemical Test Facility, the Life Sciences Test Facility, and the Joint Ambient Breeze Tunnel.

The committee met with the following individuals:

COL A. Scott Estes, Commander, DPG
SGM Stanley Morton, Jr., Command Sergeant Major, DPG
Dr. Kenneth Gritton, Technical Director, WDTC
Mr. Ryan Harris, Acting Director, WDTC
Mr. Chris Johnson, Chief, WDTC Operations Division
Dr. Chris Olson, Chief, WDTC Chemical Test Division
Mr. Jeff Garcia, Acting Chief, WDTC Dissemination & Explosives Division
Dr. Doug Andersen, Chief, WDTC Life Sciences Division
Mr. Ross Rosengren, Chief, Resource Management
Mr. Bill Brown, Test Engineering & Integration Division

Data Gathering Meeting 3:
March 16, 2012 in Natick, MA

A subgroup of the committee visited the US Army Natick Soldier Research, Development, and Engineering Center in Natick, MA. The group met with representatives of the facility and heard overview presentations about the site's capabilities and activities. The group also participated in a roundtable discussion with principle investigators and senior scientists. The committee toured and met with personnel from the following programs: Human Factors Lab and Assessment, the Molecular Science & Engineering Team, ARIEM and the Doriot Chambers, and Collective Protection and Shelters. The committee met with individuals, including: Dr. Heidi Gibson, Natick Soldier Research, Development and Engineering Center, Polymer Research Chemist.

Data Gathering Meeting 4:
March 19-20, 2012 at Various Locations

March 19, Washington, DC

The committee heard a number of briefings from representatives from agencies and individuals that have programs relevant to the CBDP initiatives or capability focus areas.

Robert Kadlec, RPK Consulting LLC
Randall Long, Director, Chemical and Biological Division, Science and Technology Directorate, Department of Homeland Security
Susan Coller-Monarez, Threat Characterization and Attribution Branch Chief, Chemical and Biological Defense Division Science and Technology Directorate, Department of Homeland Security
Elizabeth George, Director of the Cooperative Threat Reduction Directorate, Defense Threat Reduction Agency, Department of Defense
Carol Linden, Principal Deputy Director, Biomedical Advanced Research and Development Authority, Office of the Assistant Secretary for Preparedness and Response, US Department of Health and Human Services
Mike Kurilla, Director, Office of BioDefense Research Affairs, National Institute of Allergy and Infectious Diseases, National Institutes of Health
Rick Jaffe, Director, Medical Countermeasures, Strategy, and Requirements, Office of Policy and Planning, Health and Human Services
Luciana Borio, Assistant Commissioner for Counterterrorism Policy and Director, Office of Counterterrorism and Emerging Threats, Food and Drug Administration

March 20, Frederick, MD

A subgroup of the committee visited US Army Medical Research Institute for Infectious Diseases (USAMRIID) at Ft. Detrick in Frederick, MD. The group met with representatives of the facility and heard overview presentations about the site's capabilities and activities. The committee also toured BSL-2, BSL-4, and ECO facilities and toured the aerobiology laboratory. During the meeting, the group also participated in roundtable discussions with the senior scientists, program managers, and management personnel.

The committee met with the following individuals:

COL Bernard DeKoning, Commander
COL Andrea Stahl, Deputy Commander
COL Brian Gentile, Director for Administration
SGM Thomas Tuttle, Sergeant Major
Dr. Leonard Smith, Acting Science Director
Dr. Connie Schmaljohn, Senior Research Scientist
Dr. Arthur Friedlander, Supervisory Research Medical Officer
Dr. Mark Dertzbaugh, Chief, Business Plans and Programs
COL Fernando Guerena, Chief, Division of Medicine
LTC Neal Woollen, Director of Safety, Security, and Biosurety
Dr. David Norwood, Chief, Diagnostic Systems Division
Dr. Jeffrey Teska, NICBR Partnership Office
Dr. Aysegul Nalca, Chief, Center for Aerobiological Sciences
Dr. Patricia Worsham, Chief, Bacteriology Division
Dr. Louise Pitt, Chief, Virology Division
Dr. Sina Bavari, Chief, Integrated Toxicology Division
Mr. James Coffman, Chief, Office of Research and Technology Applications
Ms. Caree Vander Linden, Public Affairs Office
COL Randall Rietcheck, Director, Veterinary Medicine Division
LTC Pedro Rico, Chief, Pathology Division

March 20, Aberdeen, MD

A subgroup of the committee visited US Army Medical Research Institute for Chemical Defense (USAMRICD) and Edgewood Chemical Biological Center (ECBC) at Aberdeen Proving Ground in Aberdeen, MD. The group met with representatives, took tours of the facilities, and heard overview presentations about the site's capabilities and activities.

At USAMRICD, the committee toured the laboratories and facilities for molecular modeling, chemical surety, behavioral studies, mass spectrometry and analytical chemistry, and the vivarium. The presentations and tours were led by the following people:

COL Peter Schultheiss, Commander
COL Deborah Whitmer, Deputy Commander
Dr. John Graham, Deputy to the Commander for Research
Dr. James Dillman, Chief, Research Program Office
Dr. Doug Cerasoli, Program Advisor, Molecular and Cellular Therapeutics

Dr. Lucille Lange, Program Advisor, Toxicologic Modeling
LTC Rick Probst, Deputy Chief, Research Support Division
Dr. Ben Capacio, Program Advisor, Medical Diagnostics and Forensics
Mr. Jonathon Oyler, Chemist, Team Leader for BB Area

The committee also met with program advisors and deputy program advisors, division chiefs and deputy division chiefs, and management personnel for a roundtable discussion.

At ECBC the committee toured the McNamara Life Science Building, which houses the inhalation chamber, the standoff detection team, the genomics laboratory, and BSL-3 laboratories; the Mobile Laboratory Systems; the Advanced Chemistry Laboratory, which houses Decontamination Science, the Filter Testing Laboratory, and the Synthesis Laboratory; and the Sample Receipt Facility. Following the facility tours, committee members met with the principle investigators for wrap-up discussions then returned to the original meeting room for a roundtable discussion with facility management personnel.

The presentations and tours were led by the following people:

Dr. Joseph Corriveau, Director, Research and Technology
COL Ray Compton, Military Deputy
Dr. John Carpin, Biomedical Engineer
Dr. Christopher Whalley, (Acting) Division Chief, Toxicology and Obscurants
Dr. Mary Wade, (Acting) Branch Chief, BioDefense
Mr. Richard Vanderbeek, Branch Chief, Laser Standoff Detection
Dr. Sandy Gibbons, Research Microbiologist
Dr. Carrie Poore, Team Leader, Advanced CBRNE Training Team
Mr. George Noya, Team Leader, Mobile Labs and Kits Team
Dr. Teri Lalain, Branch Chief, Decontamination Sciences
Dr. Brent Mantooth, Principle Investigator, Modeling Simulation and Analysis
Mr. Matt Shue, (Acting) Branch Chief, Decontamination Sciences
Dr. Frederick Cox, (Acting) Division Chief, CB Protection and Decontamination
Dr. Fred Berg, Division Chief, Chemical Sciences
Mr. Chris Druyor, Chemical Engineering Technician
Mr. Jerry Pfarr, Microbiologist, Chemical Operations Branch

Data Gathering Meeting 5:
April 3, 2012 in Washington, DC

The committee invited representatives from all of the DOD Combatant Commands and the Services to brief the group on their perspectives on the CBD Program and its objectives and capabilities. At its fifth meeting, the committee held discussions with representatives from US European Command and US Northern Command. The committee also spoke with US DOD Service representatives to the Joint Requirements Office for Chemical, Biological, Radiological, and Nuclear Defense from the US Army and US Marines. Due to unforeseen circumstances, US Strategic Command and the Service representative from the US Air Force were unable to participate.

At this meeting, the committee also held roundtable discussions with program managers from the Joint Science Technology Office and the Joint Program Executive Office for Chemical and Biological Defense.

The committee was also briefed by Dr. Julie Pavlin, Deputy Director, Armed Forces Health Surveillance Center (AFHSC) and Acting Chief, Global Emerging Infections Surveillance & Response System (GEIS), on GEIS and its activities.

Data Gathering Meeting 6:
April 18, 2012 in Washington, DC

The committee held discussions with representatives from US Special Operations Command on the CBD Program and its objectives and capabilities.

Appendix B

Additional Thoughts on the Nature of the Chemical Threat

HARD INTELLIGENCE IS DIFFICULT TO OBTAIN

Considering the rather amorphous, shifting scope of the threat, the definition of what makes a "credible adversary" has changed significantly since the end of the Cold War, and as a result, the scope of required intelligence gathering efforts for both conventional and CB weapons has broadened. When considering CBW in particular, a wide range of possible weapons are available to a potential adversary, and though their preferred weapon will depend on their objectives and available capabilities and resources, this range makes collecting "hard" intelligence—e.g., well-defined intent, and capabilities with respect to the development, acquisition, and delivery mechanisms—difficult to obtain and corroborate. Though techniques for information acquisition and analysis continue to advance, the range and variability of possible CB weapons mean that the difficulties posed by this complexity will continue to make reliable intelligence gathering a challenge into the foreseeable future. Thus, it is fair to say that the CBD program cannot rely on breakthroughs in intelligence on adversaries' CB terrorism or warfare programs to determine the prioritization of its investments or deliverables.

The presence of US forces in numerous geographic niches, the diversity of potential biological threat agents, and the forces' proximity to naturally occurring diseases, makes comprehensive force health protection daunting. Additionally, the scalability of impact of naturally occurring and nefarious attacks must be addressed, i.e., not only can the massive release of a known or previously unanticipated agent have major impact on warfighter readiness and operational effectiveness but so can a small amount artfully or serendipitously focused and delivered.

What should be the strategies that underlie the CBD program? It may be that it is to defend massed US ground forces against a Soviet-like attack, but that objective is a very limited one, and current efforts—based on suits and masks of uncertain value—are focused on a historical threat, and do nothing to reduce the possibility of strategic surprise. There are so many ways that new weapons (e.g., a "chemical suicide bomber," or, in a few years, "swarm" attacks using CB weapons) can be used that fixating on the cold-war threat is probably addressing a low-risk event. Staying with a historical threat, and not rethinking the problem, is, of course, choosing: "Not to choose is to choose."

THE USE CASES HAVE CHANGED

Conventional Military Engagements

- **Use against Troops in the Field.** The evaluation of conventional military agents against equipped, protected troops in maneuver warfare (in the imagined Fulda gap battlefield) is believed to be relatively ineffective, at least in part because covering a significant area with an effective concentration requires very large amounts of agent. The correctness of this evaluation has never been tested, especially for combat in hot climates, in urban or jungle warfare, in innovative attacks against high-value facilities, or in special operations. Requiring troops to perform at high tempo in hot climates, in protective gear, would probably require much lower amounts to be effective than in cooler climates. The influence of protective gear on vision, and on the ability to work in warm climates are well understood intellectually, but their impact on the ability to perform combat operations has not been convincingly evaluated.
- **Use against Bases or High-Value Sites.** When valuable, and mission-critical, supplies are assembled in concentrated temporary storage in one place (as a port of debarkation or embarkation, a large logistics base), the use of a highly toxic and persistent agent is a plausible way of slowing or stopping operations. Especially in the early stages of a forced entry, a counterattack (perhaps a swarming attack combining rockets, clouds, and suicide missions with trucks and boats) could dramatically slow the tempo of operations (but would again require extensive preparation and synthesis, and large quantities of materials).

Unconventional Uses

Chemical weapons are probably best suited for use against targets having a small footprint or volume, since the quantity of material required may then be small.

- **Use in Forced Entry, or Special Operations.** A small group forcing entry, or localized but hidden (special operations), or restricted to a firebase, are vulnerable to local attack using chemical weapons, and hindered in their mission if present in protective gear.
- **Use by Insurgents.** An innovative group of insurgents could readily find uses for chemical weapons in denying entry to buildings or neighborhoods, in slowing operations, in disabling guards around facilities, and in increasing the lethality of IEDs by combining them with (separate) explosive attacks.
- **Use against Civilian Populations or Non-Combatants.** When it is useful to cause panic in a civilian population (to slow US military traffic through a city, to flood highways important for logistics movements with civilians, and so on) chemical weapons are attractive; relatively small quantities would be required to cause panic, and the effect on population movement would be large.
- **Use against Politically Sensitive Targets: Embassies and Missions Corporations.** In conflicts (probably *most* conflicts for the future) where political impact is more important than numbers of casualties, chemical weapons could be very effective. As an example, an attack on an embassy, other diplomatic establishments, a corporate headquarters, or an oil-transshipment facility using persistent nerve agents would attract more attention—especially to the resulting casualties—than would an attack using explosives.
- **Amplification of Effectiveness of Chemical Weapons using Social Media.** Many people fear "chemicals" to a degree that is disproportionate to the harm they might cause, and the combination of rumor and casualties could cause substantial disruption. For example, a coordinated set of attacks on subway systems would (at least in the United States, and most developed countries, if history is a guide) massively disrupt the economic performance of a surrounding city, and be both immediately expensive, and would be even more expensive later as protective measure and regulations (in the manner of 9/11) was installed. In a non-US city, panic or anger about casualties and disruption could be used to deny US access to local facilities.

CHARACTERISTICS OF CHEMICAL WEAPONS

- **Traditional Chemical Weapons.** The majority of chemical weapons were developed in the period spanning WWI and WWII. Although simple agents (e.g., phosgene, chlorine, mustard) were effective against stationary, poorly protected troops in WWI, they were almost not used in the interwar period or in WWII (Italy in Ethiopia being a counterexample), and were judged by the United States to be inefficient for land warfare in the hypothetical conflict with the Soviet Union. The Soviet Union, however, reached a different conclusion, and continued to develop both chemical and biological weapons. One possible use was considered to be as a tactical weapon to slow the tempo of operations of an opponent in maneuver warfare; a possible second use was to be in combination with biological weapons as a method of attacking survivors and remaining industrial capability after the physical destruction of cities.
- **Evolved Chemical Weapons.** The chemical weapons now of greatest concern are nerve agents and mustards, with a number of other agents—some not originally considered as weapons—also of interest and concern. The nerve agents have been highly developed, in a substantial variety of forms, with some having problematic characteristics for current equipment. Both nerve and mustard agents have the characteristics that survivors of chemical injury can require prolonged and expensive care, and thus may place a burden and expense on the force supporting them.
- **Advanced Weapons.** There are a number of newly considered nerve agents that have characteristics that require rethinking, both in terms of treaty restrictions, surveillance, detection, and protective gear. These agents *are* problematic, but we know about them, and we know their structures. Potentially as problematic are compounds that have *not* been considered (or not *yet* been considered) as weapons. The history of the pharmaceutical industry is full of compounds that are highly toxic, and design parameters for a new weapon are easily imagined (pick an essential receptor present in low concentration and antagonize it; pick an organ whose damage is life threatening or incapacitating—e.g., lung, heart, retina, pancreas—and develop a drug toxic to that system). It is worrisome that we still do not know/understand all that the Russians were doing, although we know that they had—or claimed to have—interests in a number of types of compounds that we had not actively developed in our programs in chemical weapons. Further, biology has progressed so extraordinarily

rapidly in the last three decades that one can imagine rational programs leading to quite new toxic activities.

- **Industrial Chemicals and Derivatives of Them.** For terrorist use, industrial chemicals (e.g., chlorine, phosgene, hydrogen peroxide, hydrofluoric acid) might be attractive since they are widely available, and are often shipped thorough, or in the vicinity, of cities in tank car or tank truck quantities. Although such shipments are easy to track in the developed world, they are not in the developing world, and can be adapted as weapons (as ammonium nitrate—a common fertilizer, has been adapted as an explosive). In open spaces these chemicals tend to dissipate by mixing with the atmosphere; in enclosed spaces they are more effective.
- **Weapons for Use by Terrorists.** Chemical weapons are very well suited for attacks in which the target is a "soft" biological target in an enclosed space (e.g., commuters in a subway or bus, children in schools, passengers in airplanes). HCN, H_2S are both readily prepared, and quite capable of causing a significant number of casualties. These compounds could be used to disrupt transportation systems. Benzene and carbon tetrachloride are readily available, and although not very toxic on single exposure, potential tools to cause panic since both are known to cause cancer; aflatoxin is a fungal product which is a very potent carcinogen.
- **Biological Toxin Weapons.** Biology produces a number of very toxic molecules (botulism toxin, ricin, many others: for example, peptides that alter mood, or produce fear, or interfere with judgment or memory or immune function or reproductive performance). These compounds are not volatile, and would probably have to be delivered in an aerosol. The technology of biological toxins is sophisticated, but well understood. An important feature of these materials is that the onset of symptoms can sometimes be delayed, so warning through development of symptoms may not happen until well after delivery of a complete dose. They also have the property that they fall "between" chemical and biological weapons, and are thus ambiguous in who is responsible for them.

GENERAL CONCERNS

- **Cost Effectiveness.** Chemical weapons have the potential to be effective in confined spaces, particularly when the primary objective is to cause disruption rather than large number of casualties. They are, therefore, effective as weapons in terrorism and insurgency, and against specific, localized, military targets; they are probably less effective against large-area targets, where large industrial facilities and transportation systems are required to manufacture and move the required materials.

The current programs in chemical defense in the DoD are focused on cold-war programs, and these may be concerned with the wrong threat, or perhaps a threat with a low probability relative to threats growing from attacks on the United States through low-intensity conflict intended to achieve its ends by causing popular dissatisfaction, politically unsupportable levels of casualties, and unacceptable expense.

- **Scaling to Bulk Production.** The production of chemical agents is not difficult technically (relative to either biological and nuclear weapons), but obviously requires care if the operators of the processes are not to kill themselves and their immediate neighbors. Any country capable to a moderate level of industrial activity (for example, Iraq, Iran, N. Korea, Libya, etc.) can make them, and do so in bulk. Terrorist and insurgents apparently have not been able to make the more advanced agents (or the safer but more technically sophisticated binary weapons), or have not chosen to do so.
- **Medical Treatment and Sequella.** Very little is known about the long-term sequella in human health resulting from exposure to chemical agents. Agents developed early in the history of this class of weapons (phosgene, chlorine) damaged and killed tissue, but otherwise seemed not to have hidden effects. The nerve agents have the reputation of paralyzing muscle by blocking the activity of acetyl cholinesterase, but also clearly influence this and related enzymes in other tissues, and especially in the brain; this kind of activity is presumed to be the basis for the seizures that result from exposure to nerve agents. The nature of damage to the brain and central nervous systems, and to other tissues that use acetylcholinesterase, or that react with organophosphates, is not well understood, nor is its duration or long-term consequences. If neurological damage is severe, prolonged, and expensive, the long-term care of exposed populations—both military and civilian—needs to be examined and optimized to avoid ruinous expense.
- **Innovation in Chemical Weapons.** There has been little innovation by insurgents or terrorists in the development of chemical weapons, but that fact should be only cold comfort. Explosives are more familiar, and weapons based on explosives (IEDs, EFPs, suicide bombers) have been effective, very cost effective and innovative. Chemical weapons (even simple one) provide an unfamiliar and somewhat more difficult technical barrier to

entry, but when this barrier is breached, there is great potential for development of effective, targeted uses. (This area would benefit from imaginative red-team development/thinking on the part of the United States, to avoid strategic surprise.)

- **Role of Pharmaceutical Companies.** In the course of reducing toxicity in biological chemical entities, the pharmaceutical industry is probably the greatest source of expertise on the toxicity of new, and new *classes* of, chemicals. The agricultural industries concerned with animal health, insecticides, and similar matters is another source of relevant expertise. Countries that have endogenous, developed pharmaceutical companies or industries, or have important farming sectors, are candidates for concern as sources both of skilled personnel, and as the sources for the invention or synthesis/manufacturing of chemical agents.
- **Science Base for Understanding the Effects of Chemical Weapons.** Although nerve agents, in specific, are recognized as the most important single class of chemical agents, it is remarkable that so little fundamental scientific research has been devoted to understanding how and where they act. This information is not an academic curiosity: it is *required* for the development of rational therapies for treating injuries resulting from exposure to nerve agents, and for predicting the direction of development of future nerve agents. It will also be important in recruiting the pharmaceutical industry into collaborative work (acetyl cholinesterase inhibitors are being explored, for example, as drugs for use against Alzheimer's disease).

Appendix C

Frameworks of the CBDP Enterprise Elements

JRO-CBRND: Operational Elements and Functional Capabilities.

Joint CBRN Defense Concept

• **SHAPE** – Provides the ability to characterize the CBRN hazard to the force commander – develop a clear understanding of the current and predicted CBRN situation; collect and assimilate info from sensors, intelligence, medical, etc., in near real time to inform personnel, provide actual and potential impacts of CBRN hazards; envision critical SENSE, SHIELD and SUSTAIN end states (preparation for operations); visualize the sequence of events that moves the force from its current state to those end states.

• **SUSTAIN** – The ability to conduct decontamination and medical actions that enable the quick restoration of combat power, maintain/recover essential functions that are free from the effects of CBRN hazards, and facilitate the return to pre-incident operational capability as soon as possible.

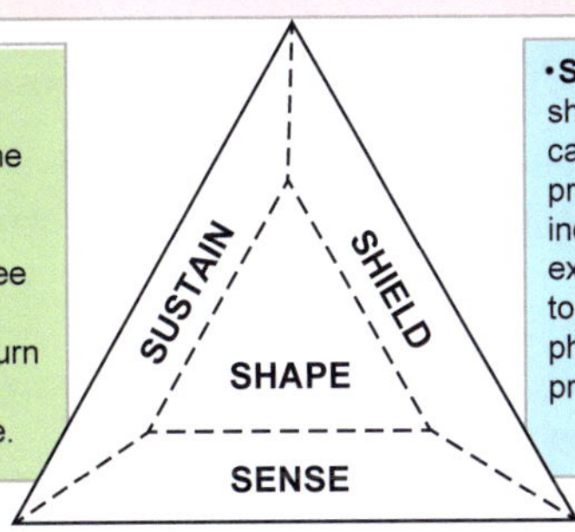

• **SHIELD** – The capability to shield the force from harm caused by CBRN hazards by preventing or reducing individual and collective exposures, applying prophylaxis to prevent or mitigate negative physiological effects, and protecting critical equipment.

• **SENSE** – The capability to continually provide the information about the CBRN situation at a time and place by detecting, identifying, and quantifying CBRN hazards in air, water, on land, on personnel, equipment or facilities. This capability includes detecting, identifying, and quantifying those CBRN hazards in all physical states (solid, liquid, gas).

JRO-CBRND: Operational Elements and Functional Capabilities.

Sense
- Biological Point Detection
- Biological Standoff Detection
- CBRN Reconnaissance
- Chemical Point Detection
- Chemical Standoff Detection
- Field Analytics
- Medical Diagnostics
- Radiological Point Detection
- Radiological Standoff Detection

Shape
- Operating Environment Analysis
- Operating Environment Mgmt Systems
- Integrated Early Warning
- Medical Surveillance
- Methods of Control

Sustain
- Biological Therapeutics
- Chemical Therapeutics
- Equipment Contamination Mitigation
- Fixed Site Contamination Mitigation
- Hazardous Waste Control
- Personnel Contamination Mitigation
- Radiological Therapeutics
- Remains Disposition

Shield
- Biological Prophylaxis
- Respiratory and Ocular Protection
- Chemical Prophylaxis
- Expeditionary Collective Protection
- Fixed Site Collective Protection
- Percutaneous Protection
- Radiological Prophylaxis

JPEO-CBD: Product Acquisition Areas

- Joint Program Manager (JPM) Information Systems (JPM-IS)
- JPM Nuclear, Biological, and Chemical Contamination Avoidance (JPM-NBC CA)
- JPM Biological Defense (JPM-BD)
- JPM Chemical and Biological Medical Systems (JPM-CBMS)
- JPM Transformational Medical Technologies
- JPM Individual Protection (JPM-IP)
- JPM Collective Protection (JPM-ColPro)
- JPM Decontamination (JPM-Decon)
- JPM Guardian (JPM-Guardian)

JSTO-CBD: Strategic Thrusts and Supporting Enablers

Disease Surveillance, Threat Detection and Point of Need Diagnostics

- Broad-Spectrum Detection
- Fieldable Dx Sequencing
- Molecular Recognition
- Host Response
- Exposure Prediction
- Functional Consequences

Adaptive Medical Countermeasures and Technologies

- Vaccines
- Immune Modulators
- Bio-Prophylaxes
- Bio-Therapeutics
- Regulatory Sciences
- Mfg Technologies

Threat Activity Sensing and Reporting

- Point Detection
- Agent Characterization
- Mathematical Recognition
- Transport & Dispersion
- Risk-Based Hazard Plots
- Agent Fate

Rapid Response and Restoration Science and Technology

- Individual Protection
- Dynamic Adaptive Materials
- Simulation and Analysis
- Rapid Prototyping
- Decontamination

Enablers

- Novel Threat Research
- Applied Math Tools
- Multifunctional Materials
- Flexible Design & Manufacturing
- Systems Biology

Appendix D

Relationship Comparison of CBDP Enterprise Frameworks to the Committee's S&T Capability Categories

DASD(CBD) Capability Focus Areas:	JRO-CBRND and JPEO-CBD Operational Elements:	JSTO-CBD Strategic Thrusts and Enablers
Surveillance	**Sense**	**Disease Surveillance, Threat Detection and Point of Need Diagnostics**
Medical Countermeasures	**Shape**	**Adaptive Medical Countermeasures and Technologies**
Hazard Mitigation	**Shield**	**Threat Activity Sensing and Reporting**
Enabling Technologies	**Sustain**	**Rapid Response and Restoration Science and Technology**
		Supportive Enablers

1. **Enabling CBRN Intelligence, Surveillance, and Reconnaissance:**

DASD(CBD)	JRO-CBRND and JPEO-CBD	JSTO-CBD
Agent Characterization and Identification	CBRN Reconnaissance	Mathematical Recognition
Physicochemical Agent Characterization	Medical Surveillance	Agent Characterization
	Integrated Early Warning	

2. Chemical and Biological Detection:

DASD(CBD)	JRO-CBRND and JPEO-CBD	JSTO-CBD
Detection (Biological & Chemical)	Chemical Point Detection	Broad Spectrum Detection
	Chemical Standoff Detection	Molecular Recognition
	Biological Point Detection	Point Detection
	Biological Standoff Detection	

3. Individual and Collective Protection:

DASD(CBD)	JRO-CBRND and JPEO-CBD	JSTO-CBD
Individual Protection	Percutaneous Protection	Individual Protection
Collective Personnel Protection	Fixed Site Collective Protection	Dynamic Adaptive Materials
Infrastructure Protection	Expeditionary Collective Protection	
	Respiratory and Ocular Protection	

4. Medical Countermeasures:

DASD(CBD)	JRO-CBRND and JPEO-CBD	JSTO-CBD
Diagnostics	Chemical Prophylaxis	Fieldable Diagnostic Sequencing
Pretreatments	Biological Prophylaxis	Host Response
Prophylaxis	Chemical Therapeutics	Vaccines
Therapeutics	Biological Therapeutics	Immune Modulators
Toxicological Agent Characterization	Medical Diagnostics	Bio-prophylaxis
Pathophysiological Response		Bio-therapeutics
		Regulatory Science
		(Medical) Manufacturing Technologies

5. Hazard Assessment, Management and Decontamination:

DASD(CBD)	JRO-CBRND and JPEO-CBD	JSTO-CBD
Information Management	Methods of Control	Exposure Prediction
Individual Decontamination	Hazardous Waste Control	Functional Consequences
Environmental Decontamination	Personnel Containment Mitigation	Transport & Dispersion
Human Remains Decontamination	Equipment Containment Mitigation	Risk-Based Hazard Plots
Hazardous Exposure Containment	Fixed-Site Containment Mitigation	Agent Fate
	Remains Disposition	Decontamination

6. Cross-Cutting Science and Technology:

DASD(CBD)	JRO-CBRND and JPEO-CBD	JSTO-CBD
Environmental Agent Fate	Field Analytics	Simulation and Analysis
Agent Dissemination & Modeling	or Operating Environment Management Systems	Rapid Prototyping
Simulant Development	or Operating Environment Analysis	Novel Threat Research
		Applied Math Tools
		Multifunctional Materials
		Flexible Design and Manufacturing
		Systems Biology

Appendix E

Committee Member Biographies

Dr. Miriam E. John (*chair*) is serving in various consulting and board roles since her retirement as the vice president of Sandia National Laboratory in Livermore, California. During her Sandia career, she worked on a wide variety of programs, including nuclear weapons, chemical and biological defense, missile defense, and solar energy, and provided leadership for a number of the laboratory's energy, national security and homeland security programs. She is a member of the DoD's Defense Science Board and Threat Reduction Advisory Committee. She is also the chair of the National Research Council's Naval Studies Board and serves on the board of directors of the National Institute for Hometown Security. She is a past member of the Air Force Scientific Advisory Board, the Board on Army Science and Technology, and DOE's National Commission on Science and Security. She was appointed a national associate of the National Academies of Science and Engineering. She chairs the California Council on Science and Technology. She is a member of the board of advisors for MIT Lincoln Laboratory, the board of directors for Draper Laboratory, and external advisory board of Savannah River National Laboratory. She is a member of the board of directors of SAIC and the Strategic Advisory Board for RedX Defense Systems. She is also a member of the director's review committee for the National Ignition Facility at Lawrence Livermore National Laboratory, and has served on the director search committees for both Lawrence Livermore and Los Alamos National Laboratories. She is a member of the Dean's advisory board for the School of Science and Engineering and chairs the advisory board for the Department of Chemical and Biomolecular Engineering at Tulane University, where she has been recognized as an outstanding alumna.

Dr. David R. Franz, Ph.D., D.V.M., was formerly the vice president and chief biological scientist at Midwest Research Institute in Frederick, Maryland, and senior advisor to the Office of the Assistant to the Secretary of Defense for Nuclear, Chemical and Biological Defense Programs for the Defense Threat Reduction Agency. Dr. Franz served in the U.S. Army Medical Research and Materiel Command for 23 of 27 years on active duty and retired as a colonel. He served as commander of the U.S. Army Medical Research Institute of Infectious Diseases (USAMRIID) and as deputy commander of the Medical Research and Materiel Command. Prior to joining the command, he served as group veterinarian for the 10th Special Forces Group (Airborne). Dr. Franz was the chief inspector on three United Nations Special Commission biological warfare inspection missions to Iraq and served as technical advisor on long-term monitoring. He also served as a member of the first two US-UK teams that visited Russia in support of the Trilateral Joint Statement on Biological Weapons and as a member of the Trilateral Experts' Committee for biological weapons negotiations. Dr. Franz was technical editor for the Textbook of Military Medicine on Medical Aspects of Chemical and Biological Warfare released in 1997. Current standing committee appointments include the Defense Intelligence Agency Red Team Bio-Chem 2020, the National Academy of Sciences Committee on International Security and Arms Control and the Board on Life Sciences, and the Department of Health and Human Services National Science Advisory Board for Biosecurity. He serves on the Boards of the Federation of American Scientists and Integrated Nano-Technologies. Dr. Franz holds an adjunct appointment as professor for the Department of Diagnostic Medicine and Pathobiology at the College of Veterinary Medicine, Kansas State University, and serves on the Dean's advisory council. The current focus of his activities relates to the role of international engagement in the life sciences as a component of national security policy. Dr. Franz holds a D.V.M. from Kansas State University and a Ph.D. in physiology from Baylor College of Medicine.

Ms. Jill Hruby is the Sandia National Laboratories vice president for energy, security and defense technologies. The energy, security and defense technologies organization primarily supports Sandia's mission efforts in energy and resource systems research and development, nuclear power, environmental quality, the reduction of the proliferation of weapons of mass destruction and the global threat of terrorism, and the protection of nuclear and other vital national assets. Jill will also lead Sandia's International, Homeland, and Nuclear Security Strategic Management

Unit (SMU), including Sandia's strategic initiative on nuclear security. This initiative focuses on all aspects of nuclear security including non-proliferation, technology support to arms control activity, global nuclear security and threat reduction, nuclear asset protection and detection and response to weapons of mass destruction. Most recently the director of Homeland Security and Defense Systems at Sandia's Livermore, Calif., site, she has been with Sandia for more than 25 years. She has served as Sandia's director of materials and engineering sciences, where she was responsible for materials research and development and microsystem fabrication and performance. Over the course of her Sandia career, she has also been actively engaged with nanoscience research, hydrogen storage, solar energy research, mechanical component design, thermal analysis and microfluidics. Jill is a member of the Board on Chemical Sciences and Technology of the National Academies.

Dr. Anna Johnson-Winegar served as the deputy assistant to the Secretary of Defense (Chemical and Biological Defense Programs) from 1999 until her retirement in 2003. She acted as the single focal-point within the Office of the Secretary of Defense (OSD) responsible for oversight, coordination, and integration of the chemical/biological (CB) defense, counterproliferation support, chemical demilitarization, and Assembled Chemical Weapons Assessment (ACWA) programs. She represented the Department of Defense on multiple interagency and international groups addressing chemical and biological issues. She provided Congressional testimony on numerous occasions during this time. She also participated as a biological weapons inspector in Iraq for the United Nations under UNSCOM. In 1998, she received the lifetime achievement award from Women in Science and Engineering. Upon her retirement from civil service, she received the Department of Defense Meritorious Service Award (with bronze palm), Presidential Rank Award as a Meritorious Executive in the Senior Executive Service, the gold medal from the National Defense Industrial Association, and numerous other recognitions. In 2006 she received the Distinguished Alumna Award from Hood College, her alma mater. She currently is engaged in private consulting work for industry, academic, and government clients.

Dr. Charles E. Kolb is the president and chief executive officer of Aerodyne Research, Inc. He joined Aerodyne as a senior research scientist in 1971. At Aerodyne, his personal areas of research have included atmospheric and environmental chemistry, combustion chemistry, and the chemical physics of rocket and aircraft exhaust plumes. In the area of atmospheric and environmental chemistry, Dr. Kolb initiated Aerodyne's programs to develop and utilize tunable infrared laser spectrometers and aerosol mass spectrometers for the identification and quantification of sources, sinks and ambient concentration distributions of trace atmospheric gases and aerosol particles involved in urban, regional and global pollution problems, as well as the development of spectral sensing techniques to quantify soil pollutants. He received a B.S. in chemistry from MIT and a M.S. and Ph.D. in physical chemistry from Princeton University. He is a fellow of the American Chemical Society, the American Physical Society, the American Geophysical Union, the Optical Society of America and the American Association for the Advancement of Science. Dr. Kolb is also a member of the Board on Chemical Sciences and Technology of the National Academies and a National Associate of the National Academies.

Dr. C. Rick Lyons was named director of the Infectious Disease Research Center at Colorado State University in 2010. Dr. Lyons is a physician scientist trained as a hematologist/oncologist. He received his M.D. and doctorate from the University of Texas Southwestern Medical School in Dallas, Texas. He received his Ph.D. in immunology and did his training in hematology/oncology at the Brigham and Women's Hospital in Boston, Massachusetts. He comes to Colorado State University from the University of New Mexico Health Science Center in Albuquerque where he was professor of medicine and director of the Center for Infectious Diseases and Immunology. His scientific expertise is in developing animal models of human diseases that can be used to translate products into humans. Dr. Lyons has over twenty five years experience in developing and performing research in animal models of infectious disease. There are three main emphases in his research: 1) Develop the most accurate animal models of infection that mimic human disease; 2) Apply cutting edge technology to analyze the endpoints during in vivo infection; and 3) Develop strong collaborations with internal and external investigators to bring the most expertise to bear on these issues. In the last ten years he has focused his research on a variety of emerging infections particularly in the field of bioweapons including *Bacillus anthracis* and *Francisella tularensis* using a variety of species to examine their pathogenesis including mice, rats, rabbits and primates.

Dr. Jon Mogford is the associate vice chancellor for strategic initiatives with the Texas A&M University System. He entered this role after serving over 6 years at the Defense Advanced Research Projects Agency (DARPA), where he was a program manager then deputy director (2010-2011) of the Defense Sciences Office (DSO). DSO's research portfolio spans from fundamental science to applications by identifying and pursuing activity within the science and engineering research communities and transforming these ideas into new DoD capabilities in the physical sciences, training/human effectiveness, biological warfare defense, materials, mathematics and biology. His DARPA programs included scar-free wound regeneration, metabolic control strategies for hemorrhagic shock, biomarker-responsive biomaterials for drug delivery, stem cell-based bioreactor production of universal donor red blood cells, computational design of proteins, and "wound stasis" biomaterials. Dr. Mogford obtained his bachelor's degree in zoology from Texas A&M University and doctorate in medical physiology from the Texas A&M University Health Science Center. He continued research in vascular physiology at the University of Chicago as a postdoctoral fellow from 1997-98 then transitioned to the field of wound healing at Northwestern University both as a research associate and a research assistant professor from 1998-2003. He has authored or co-authored 29 peer-reviewed publications.

Dr. Randall S. Murch is an associate director at the Center for Technology Security and Policy and professor in practice at Virginia Polytechnic Institute and State University, National Capital Region, Alexandria, Virginia. He is also a visiting professor, Program on Science and Security, Department of War Studies, King's College London, UK. Dr. Murch's first career was as a special agent to the Federal Bureau of Investigation (FBI), where he focused on counterintelligence, counterterrorism, forensic science, technology development and technical operations, and WMD terrorism. He served in the Indianapolis, Los Angeles and New York field offices, and the Laboratory, Intelligence and Technical Services Divisions. He created the FBI's WMD forensic investigative program. The WMD forensic program has since become a national priority and has been embraced by other federal agencies. He also led forensic investigative aspects of a number of major terrorism cases, and initiated a number of new programs for both the FBI Laboratory and technical investigative program. Toward the end of his career, he was detailed from the FBI to the Defense Threat Reduction Agency (DTRA), Department of Defense, where he led advanced studies on complex current and future challenges dealing with weapons of mass destruction. Before Virginia Tech, Dr. Murch was at the Institute for Defense Analyses (IDA) where he led and participated in studies for the defense, intelligence and homeland security communities. He has or is serving on several National Academies and Department of Defense boards and study committees. Dr. Murch holds B.S., M.S. and Ph.D. degrees in the life sciences.

Dr. Donald Prosnitz is a senior principal researcher (adjunct) at RAND Corporation, a visiting scholar at the physics department of the University of California, Berkeley, and an independent technical consultant. His current activities include research on free-electron lasers and a range of studies at RAND concentrating on the utilization of technology to solve national and homeland security issues. Dr. Prosnitz was previously the deputy associate director (programs) for Non-Proliferation, Homeland and International Security at Lawrence Livermore National Laboratory (LLNL) where he was responsible for overseeing all of the directorate's technical programs. He spent two years as an assistant professor at Yale University before joining LLNL. Over the next three decades, he conducted research on lasers, particle accelerators, high power microwaves, free electron lasers, and remote sensing, and managed the design, construction, and operation of numerous research facilities. In 1990, he was awarded the U.S. Particle Accelerator Award for Achievement in Accelerator Physics and Technology. In 1999, Dr. Prosnitz was named the first chief science and technology advisor for the Department of Justice (DOJ) by Attorney General Janet Reno. He was responsible for coordinating technology policy and technology projects among the DOJ's component agencies and with state and local law enforcement entities. In 2002, he was named a fellow of the American Physical Society. He is a former chair of the American Physical Society Forum on Physics and Society. He recently served on the National Research Council Committee to Review the Department of Homeland Security's Approach to Risk Analysis. Dr. Prosnitz received his B.S. from Yale University and his Ph.D. in physics from the Massachusetts Institute of Technology. He is a licensed amateur radio operator and an active member of his community's CERT (Community Emergency Response Team).

Mr. Tom Slezak has been involved with bioinformatics at Lawrence Livermore National Laboratory (LLNL) for 30 years after receiving B.S. and M.S. degrees in computer science from the University of California, Davis. Tom is currently the associate program leader for Informatics for the Global Security Program efforts at LLNL. He was involved with the Human Genome Program from 1987-2000, leading the informatics efforts at LLNL and then the DOE's Joint Genome Institute from 1997-2000. In 2000 he began to build a pathogen bioinformatics team at LLNL pioneering a novel whole-genome analysis approach to DNA signature design. His team developed signature targets for multiple human pathogens that were used at the 2002 winter Olympic Games under the BASIS program and later adapted for use nationwide in the Department of Homeland Security BioWatch program. Tom's team is currently focusing on signatures of mechanisms of virulence, antibiotic-resistance, and evidence of genetic engineering. They have been focusing on detecting novel, engineered, and advanced biothreats for several years, sponsored by multiple agencies. Tom has chaired or served on multiple advisory boards, including the rice genome project, mouse and maize genetics databases, spruce tree genome project (Canada), plant pathogens, a NIAID sequencing center contract renewal, and a CDC Blue Ribbon Panel on bioinformatics. He served on the National Biosurveillance Advisory Subcommittee and on a previous National Research Council panel on select agent science.

Dr. Henry H. Willis is a professor of policy analysis at the Pardee RAND Graduate School, the acting director of the RAND Homeland Security and Defense Center, and a senior policy researcher at RAND Corporation. His research has applied risk analysis tools to the study of public policy in the areas of counterterrorism, disaster response and emergency preparedness, public health and safety, energy systems, and environmental pollution control. He is the author of dozens of publications, book chapters and op-ed pieces and has testified before Congress as an expert on applying risk analysis to terrorism security policy. Dr. Willis' recent research has involved evaluating emergency preparedness programs like the Cities Readiness Initiative and analyzing the effectiveness of nuclear detection technologies used to improve supply chain security. He serves on the editorial board of the journal Risk Analysis. Dr. Willis earned his Ph.D. degree from the Department of Engineering and Public Policy at Carnegie Mellon University and holds degrees in chemistry and environmental studies from the University of Pennsylvania (B.A.) and in environmental science from the University of Cincinnati (M.A.).

CONSULTANTS TO THE COMMITTEE:

Dr. Patrick J. Scannon is company founder, executive vice president, chief scientific officer and a member of the board of directors for XOMA, a biopharmaceutical company developing novel therapeutics based on recombinant human monoclonal antibody technologies. Since 1980, Dr. Scannon has directed the company's product identification, evaluation and clinical testing programs. As chief scientific officer, he currently heads the company's preclinical product discovery programs in the areas of metabolic, ophthalmic, inflammatory and oncologic diseases. Dr. Scannon holds a Ph.D. in organic chemistry from the University of California, Berkeley and a medical degree from the Medical College of Georgia. A board-certified internist, he completed his medical internship and residency in internal medicine at the Letterman Army Medical Center in San Francisco, achieving the rank of major in the U.S. Army Medical Corps. Dr. Scannon is the inventor or co-inventor of several issued U.S. and international patents, and has published numerous scientific and clinical abstracts and papers. Dr. Scannon currently is a member of the National Biodefense Science Board (NBSB), a federal advisory board for the Department of Health and Human Services and a member of the Defense Sciences Research Council

(DSRC), a research board for Defense Advanced Research Projects Agency (DARPA). He has served as a member for the Threat Reduction Advisory Committee for the Department of Defense, chairing the ChemBiowarfare Defense Panel and also has been a participant in the Biodefense Network Assessment for the Department of Homeland Security. Dr. Scannon has served as a trustee of the University of California Berkeley Foundation and as a member of the University of California Berkeley Chancellor's Community advisory board. He has served or is serving on the boards of several companies and institutions.

Dr. George M. Whitesides received an A.B. degree from Harvard University in 1960 and a Ph.D. from the California Institute of Technology (with J.D. Roberts) in 1964. He was a member of the faculty of the Massachusetts Institute of Technology from 1963 to 1982. He joined the Department of Chemistry of Harvard University in 1982, and was department chairman 1986-1989, and Mallinckrodt Professor of Chemistry from 1982-2004. He is now the Woodford L. and Ann A. Flowers university professor.